# PyTorch
# 深度学习入门与实战

王宇龙 编著

中国铁道出版社有限公司
CHINA RAILWAY PUBLISHING HOUSE CO., LTD.

## 内 容 简 介

书中以案例形式详细介绍了 PyTorch 的各种实战应用。具体内容包括 PyTorch 与 TensorFlow 的对比和 PyTorch 的发展现状，张量 Tensor 和自动微分 Autograd 及其具体应用，PyTorch 构建神经网络，基于 PyTorch 构建复杂应用，PyTorch 高级技巧与实战应用，网络剪枝应用。

本书内容由浅入深，适合深度学习的初学者阅读学习，可帮助机器学习、计算机学科相关专业的学生或从业人员快速掌握 PyTorch。

**图书在版编目（C I P）数据**

PyTorch 深度学习入门与实战/王宇龙编著. —北京：中国铁道出版社有限公司，2020.9
　ISBN 978-7-113-27004-9

　Ⅰ.①P… Ⅱ.①王… Ⅲ.①机器学习 Ⅳ.①TP181

中国版本图书馆 CIP 数据核字(2020)第 104511 号

| 书　　名：PyTorch 深度学习入门与实战 |
|---|
| 　　　　　PyTorch SHENDUXUEXI RUMEN YU SHIZHAN |
| 作　　者：王宇龙 |

| 责任编辑：于先军 | 读者热线：（010）63560056 |
|---|---|
| 责任印制：赵星辰 | 封面设计：MXK DESIGN STUDIO |

| 出版发行：中国铁道出版社有限公司（100054，北京市西城区右安门西街 8 号） |
|---|
| 印　　刷：中国铁道出版社印刷厂 |
| 版　　次：2020 年 9 月第 1 版　2020 年 9 月第 1 次印刷 |
| 开　　本：787 mm×1 092 mm　1/16　印张：13.5　字数：265 千 |
| 书　　号：ISBN 978-7-113-27004-9 |
| 定　　价：69.80 元 |

**版权所有　侵权必究**

凡购买铁道版图书，如有印制质量问题，请与本社读者服务部联系调换。电话：（010）51873174
打击盗版举报电话：（010）51873659

配套资源下载地址：

链接：https://pan.baidu.com/s/1nW3YoupN2RruLeH-L7A76g
提取码：0o1t

http://www.m.crphdm.com/2020/0630/14270.shtml

# 前　言

深度学习作为近些年来人工智能领域发展非常迅速，应用普及范围非常广的技术，受到越来越多的关注。其在诸多领域创造出的许多成果，让人们开始畅想未来的人工智能时代。因此无论是科研人员还是相关领域从业者，都在密切关注并学习了解这一领域的进展。

但是回想在十年前，想要让一个初学者迅速进行一个训练深度学习模型的实验则是十分困难的。归根结底是因为当时并没有一款广泛普及的深度学习编程框架。每个研究者需要编写出复杂的反向传播求导过程，处理数据多线程加载，GPU 并行计算等等技术细节。这其实也限制了当时初期深度学习技术的普及。但是好在随着如 Caffe、Theano、TensorFlow、PyTorch 等等一系列深度学习框架的出现，让普通的研究者也可以轻松实现大规模数据并行编程，这才使得深度学习领域发展百花齐放，日益蓬勃。

本书所介绍的便是以上诸多编程框架中，我个人认为当前使用方法较灵活易懂的 PyTorch 可微编程框架。虽然 PyTorch 发展才不过两三年时间，但是凭借其良好的应用接口，简洁直观的编程模型，动态灵活的计算过程，成为后起之秀，逐渐被广大研究者所推崇使用。我也是在平时科研期间主要应用 PyTorch 框架，书中介绍的诸多内容也是我平时的经验总结。可以略夸张地说，在使用 PyTorch 框架编程之后，限制你科研成果进展的只剩下你的想象力和创造力，在编程实现上几乎没有任何阻碍了。

全书分为 6 章。第 1 章是 PyTorch 简介，可以让读者简略了解 PyTorch 框架的发展起源历程。第 2~5 章则由浅入深逐步介绍 PyTorch 的使用方法和编程技巧。其中先介绍了 PyTorch 作为支撑可微编程模型的基础张量以及自动求导，然后介绍高层封装模块用以搭建神经网络，接着讲解实际的应用任务实现，这里需要涉及到诸多外延内容，如数据加载预处理、模型加载保存、多 GPU 分布式计算、自定义扩展实现底层算子等等。最后一章通过一个完整的实验任务全面展示 PyTorch 在实际科研中的应用。在每一章讲解过程中，都会附带一些实战应用案例，以便于读者能够快速应用相关知识解决实际问题。这些实际问题我在设计之时，特别考虑了其实用性和理解性之间的平衡。既可以起到应用相关知识技能点，又不使其沦为过于简单的 Toy Example。同时许多实战应用和案例讲解均是以当前深度学习领域前沿的研究成果进行展示，让读者在学习 PyTorch 框架之时，并不局限于只学习技术，更能了解跟踪当前科研领域的最新进展。

本书是我编写的第一部书籍，在写作过程中也存在着诸多不足。比如全书聚焦于

PyTorch 应用环节，对于其框架底层具体实现讲解缺乏。同时限于篇幅和内容范围，许多案例讲解我只能以简化设置形式进行讲解。对于实际应用中有着更为复杂处理流程的大型任务，本文缺少进一步展示。虽然如此，我仍满怀诚意尽我所能使本书完美。我希望这本书可以成为人工智能领域初学者的启发，促使读者能够大胆探索当前人工智能各个方向的研究领域，创造出属于自己的研究成果。最后欢迎各位批评指正，感谢成书过程中所有帮助过我的人！

作　者

2020 年 7 月

# 目　录

## 第 1 章　PyTorch 简介

1.1　深度学习简介 ................................................................................................. 1
1.2　PyTorch 的由来 ............................................................................................. 2
　　1.2.1　深度学习框架回顾 ............................................................................. 2
　　1.2.2　PyTorch 前身：Torch7 ....................................................................... 4
　　1.2.3　Torch7 的重生 ..................................................................................... 5
1.3　PyTorch 与 TensorFlow 对比 ....................................................................... 5
　　1.3.1　TensorFlow 简介 ................................................................................. 6
　　1.3.2　动静之争 ............................................................................................. 6
　　1.3.3　二者借鉴融合 ..................................................................................... 7
　　1.3.4　PyTorch 的优势 ................................................................................... 7
1.4　PyTorch 发展现状 ......................................................................................... 8
　　1.4.1　主要版本特点回顾 ............................................................................. 8
　　1.4.2　准备工作 ............................................................................................. 8

## 第 2 章　PyTorch 基础计算

2.1　PyTorch 核心基础概念：张量 Tensor ..................................................... 11
　　2.1.1　Tensor 基本介绍 ............................................................................... 11
　　2.1.2　Tensor 数学运算操作 ....................................................................... 15
　　2.1.3　Tensor 索引分片合并变换操作 ....................................................... 20
　　2.1.4　Tensor 类成员方法 ........................................................................... 22
　　2.1.5　在 GPU 上计算 ................................................................................. 24
2.2　PyTorch 可微编程核心：自动微分 Autograd ......................................... 25
　　2.2.1　PyTorch 自动微分简介 ..................................................................... 25
　　2.2.2　可微分张量 ....................................................................................... 25
　　2.2.3　利用自动微分求梯度 ....................................................................... 26
　　2.2.4　Function：自动微分实现基础 ......................................................... 29
　　2.2.5　注意事项 ........................................................................................... 31
2.3　PyTorch 应用实战一：实现卷积操作 ..................................................... 34
　　2.3.1　卷积操作 ........................................................................................... 34
　　2.3.2　利用张量操作实现卷积 ................................................................... 36
2.4　PyTorch 应用实战二：实现卷积神经网络进行图像分类 ..................... 38

# 第 3 章　PyTorch 构建神经网络

- 3.1 PyTorch 神经网络计算核心：torch.nn .................................................. 43
  - 3.1.1 nn.Module 概述 .................................................................................. 43
  - 3.1.2 结构化构建神经网络 .......................................................................... 47
  - 3.1.3 经典神经网络层介绍 .......................................................................... 49
  - 3.1.4 函数式操作 nn.functional .................................................................. 53
- 3.2 PyTorch 优化器 .......................................................................................... 55
  - 3.2.1 torch.optim 概述 ................................................................................. 55
  - 3.2.2 经典优化器介绍 .................................................................................. 56
  - 3.2.3 学习率调整 .......................................................................................... 57
- 3.3 PyTorch 应用实战一：实现二值化神经网络 .......................................... 59
  - 3.3.1 二值化网络 BinaryNet 概述 ............................................................... 59
  - 3.3.2 具体实现 .............................................................................................. 60
- 3.4 PyTorch 应用实战二：利用 LSTM 实现文本情感分类 ........................ 63
  - 3.4.1 文本情感分类 ...................................................................................... 63
  - 3.4.2 具体实现 .............................................................................................. 65

# 第 4 章　基于 PyTorch 构建复杂应用

- 4.1 PyTorch 数据加载 ...................................................................................... 70
  - 4.1.1 数据预处理：torchvision.transforms ................................................ 70
  - 4.1.2 数据加载：torch.utils.data ................................................................. 73
- 4.2 PyTorch 模型搭建 ...................................................................................... 77
  - 4.2.1 经典模型复用与分享：torchvision.models ...................................... 78
  - 4.2.2 模型加载与保存 .................................................................................. 79
  - 4.2.3 导出为 ONNX 格式 ............................................................................ 85
- 4.3 训练过程中日志记录与可视化 ................................................................ 89
- 4.4 PyTorch 应用实战一：在 CIFAR10 数据集进行神经网络结构搜索 ...... 93
  - 4.4.1 可微分网络架构搜索 DARTS 介绍 .................................................. 94
  - 4.4.2 简化问题建模：以 ResNet 为例 ....................................................... 95
  - 4.4.3 具体实现 .............................................................................................. 96
- 4.5 PyTorch 应用实战二：在 ImageNet 数据集进行弱监督物体定位 ........ 108
  - 4.5.1 GradCAM 解释显著图方法介绍 ....................................................... 108
  - 4.5.2 弱监督物体定位任务 .......................................................................... 109
  - 4.5.3 具体实现 .............................................................................................. 110

# 第 5 章　PyTorch 高级技巧与实战应用

5.1 PyTorch 并行计算 ........................................................................................................ 118
　　5.1.1 大规模数据集加载 ........................................................................................... 118
　　5.1.2 模型的高效并行计算 ....................................................................................... 122
　　5.1.3 加速模型计算和减少显存使用 ....................................................................... 125
5.2 扩展 PyTorch ................................................................................................................. 126
　　5.2.1 利用 C++和 CUDA 实现自定义算子 ............................................................. 126
　　5.2.2 利用 TorchScript 导出 PyTorch 模型 .............................................................. 136
5.3 丰富的 PyTorch 资源介绍 ........................................................................................... 145
5.4 PyTorch 应用实战一：在 ImageNet 上训练 MobileNet-V2 网络 .............................. 146
　　5.4.1 MobileNet-V2 网络介绍 ................................................................................... 146
　　5.4.2 具体实现 ........................................................................................................... 147
5.5 PyTorch 应用实战二：利用 CUDA 扩展实现 MixConv 算子 ................................... 157
　　5.5.1 MixConv 算子介绍 ........................................................................................... 157
　　5.5.2 借鉴 Depthwise 卷积实现思路 ........................................................................ 158
　　5.5.3 具体实现 ........................................................................................................... 160

# 第 6 章　PyTorch 完整实战讲解——网络剪枝应用

6.1 网络剪枝介绍 ............................................................................................................... 169
　　6.1.1 剪枝方法分类 ................................................................................................... 169
　　6.1.2 基于权重通道重要性的结构化剪枝 ............................................................... 170
　　6.1.3 问题定义与建模 ............................................................................................... 170
6.2 具体实现思路 ............................................................................................................... 171
　　6.2.1 如何附属控制门值 ........................................................................................... 171
　　6.2.2 剪枝结构搜索 ................................................................................................... 172
　　6.2.3 剪枝模型训练 ................................................................................................... 174
6.3 完整代码实现 ............................................................................................................... 175
　　6.3.1 模型搭建 ........................................................................................................... 176
　　6.3.2 剪枝器实现 ....................................................................................................... 181
　　6.3.3 学习控制门变量 ............................................................................................... 183
　　6.3.4 剪枝模型 ........................................................................................................... 187
　　6.3.5 训练模型 ........................................................................................................... 189
　　6.3.6 规模化启动训练任务 ....................................................................................... 193
6.4 实验结果 ....................................................................................................................... 198

**参考文献**

# 第 1 章 PyTorch 简介

本章中，我们将初步介绍 PyTorch 框架的相关情况，回顾 PyTorch 的发展历程，了解为何该框架可以在一两年之内就能得到学术界及业界的青睐，并呈现出蓬勃发展的态势。了解这些情况，可以让我们更全面地认识到 PyTorch 框架的优劣，以及深度学习计算框架的发展态势，对于我们未来选择深耕的技术栈很有帮助。同时，回顾这一过程，我们也会体会到无论当前已有的生态看似多么强大，只有符合需求能解决实际应用痛点的框架平台，才会真正受到多数人的认同。

## 1.1 深度学习简介

在介绍 PyTorch 框架之前，我们有必要简单回顾一下深度学习最近十年左右的发展情况。所谓"简单"，是因为深度学习发展到今天，已经是遍地开花琳琅满目，在计算机视觉、自然语言处理、语音识别、强化学习、搜索推荐等等诸多领域，都展现出绝对的优势并被广泛应用。限于篇幅我们只能选取有代表性发展的一些事件概览深度学习的发展情况。而所谓"十年"，我认为主要指 2010 至今以来的十年，是深度学习真正从原来一个实验室算法原型层面，走向工业界大规模实际应用的转变时间段。而 PyTorch 正是在此大发展背景下所涌现出的诸多计算框架中具有代表性的一个。

其实早在 20 世纪 80 年代，深度学习的实现主体——神经网络已经开始研究。包括其模型如卷积神经网络[1]（Convolution Neural Network，CNN），反馈神经网络[2]（Recurrent Neural Network，RNN），求解方法如反向传播算法[3]（Backpropagation），均已被提出。甚至现在业界翘首以待的自动驾驶，也曾在 1985 年由 CMU 团队实现过[4]。而在 2012 年以 AlexNet[5]在 ImageNet 图像识别任务夺冠作为标志，开启了深度学习的时代。在计算机视觉领域中，诸多深度模型架构被开发出来，如 2013 年的 VGGNet[6]，2014 年的 GoogLeNet[7]，2015 年的 ResNet[8]等等。深度学习又进一步向其他领域大举进发。2016 年 Deepmind 的 AlphaGo[9]横空出世，让人们见识到了结合深度神经网络的强化学习竟有如此超越人类的能力。2017~2018 年提出的 Transformer[10]、Bert[11]等自然语言预训练模型又让人仿佛看到 ResNet 那种大巧若拙的模型拟合能力。而神经架构搜索[12]（Neural Architecture Search，NAS）的兴起，又进一步突破人们的认知，让人期待深度学习模型的能力边界究竟在何处。

纵观近十年深度学习的异军突起，笔者认为主要源于三个方面。

- **大数据爆发**：海量数据的增加使得深度学习算法可以有效地提取反复出现的特征，

使得模型泛化（Generalization）问题逐渐变成了一个依赖于数据规模的问题，即越大量的数据越有可能覆盖多种样本，从而实现了对于未知样本准确的预测。

- **计算力提升**：自从 GPU 的并行计算能力被发掘用以实现深度学习模型中常见的密集矩阵运算，深度学习模型的训练过程被极大的缩减。
- **可微分编程**（Differentiable Programming）**普及**：其实深度学习抛开神经网络这个实现主体来说，其本质在于可微分编程，即利用模型自身的可导性质，依赖于数据自发完成参数学习。这使得原本的研究范式中求解优化问题变成了计算图设计，因为一旦计算流程确定，其反向回传求导过程也被确定。因此极大地解放了建模问题的门槛限制，从而被广泛应用接受。

而能够集以上三者于一身的必然是具有处理大规模数据，有效利用 GPU 高性能计算的可微分编程框架。事实上，在深度学习崛起的过程当中，适配于深度学习计算模式的编程框架起到了至关重要的作用。PyTorch 正是这些框架中的杰出代表。接下来我们将回顾 PyTorch 的发展过程，从而可以更好地理解虽然 PyTorch 作为后起之秀，但却超越了前辈成为广受欢迎的编程框架。从图 1.1 可以看出，近几年各大人工智能会议中使用 PyTorch 框架论文所占比重呈逐年增加趋势（图片来源链接：https://chillee.github.io/pytorch-vs-tensorflow/）。

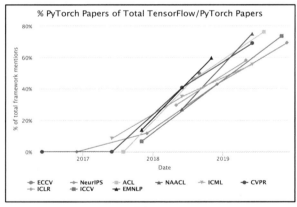

图 1.1

## 1.2 PyTorch 的由来

在本节中我们将介绍深度学习代表性框架，包括早期的 Caffe、Theano、MatConvNet 以及 PyTorch 的框架前身 Torch7。在这些回顾过程中，我们会逐渐理解当时 PyTorch 的发展背景以及设计理念。

### 1.2.1 深度学习框架回顾

之前我们谈到过，2012 年的神经信息处理大会（Neural Information Processing

Symposium，NeurIPS）上，一篇题为"Imagenet Classification with Deep Convolutional Neural Networks"[5]的论文横空出世，该论文使用了基于 CNN 的 AlexNet 模型，使得在 ImageNet 这个有着大约一百二十多万张图片的图像分类数据上，成功将 top-5 分类错误率降低到 15.3%，远超当年 ILSVRC 比赛第二名所取得的 26.2%错误率。而该文的第一作者 Alex Krizhevsky 也发布了其实现代码 cuda-convnet①，成为第一个可以利用 GPU 计算实现 CNN 的深度学习框架。然而称其为框架其实有些夸大，因为源代码组织结构松散，且其中涉及大量 CUDA 实现的基本运算核操作，混杂在一起让人头大。而 Alex 本人也并没有尽力维护推广这份代码，甚至在其毕业后远离学术界，有种"事了拂衣去，深藏功与名"的感觉。

而真正意义上被广泛接受且拥有巨大影响力，还可以同时处理大规模数据并利用 GPU 计算力的框架，当属 2013 年推出的 Caffe②。时至今日，仍有不少创业公司或业界部门在沿袭使用着"魔改"版 Caffe，可见其影响之深远。Caffe 开发来自于伯克利视觉学习中心（BVLC）团队。起初是贾扬清博士在毕业期间所编写的一套代码，但是由于其清晰可复用，并且很好地处理了底层 GPU 算子实现与上层模型搭建的解耦，以及成功地复现了 AlexNet 结果，使得学术界争相尝试。毕竟每个科研人员最头疼的便是结果复现环节，论文描述很丰满，现实实践很骨感。有了这一套框架，再也不用喝 Caffeine 熬夜赶代码了。

我们以现在的眼光，回顾 Caffe 框架，其实现的是一种宣告式编程（Declarative Programming）框架。具体来说，Caffe 需要单独编写 protobuf 文件来预先设置神经网络结构和连接情况。而一旦设定好网络之后，其运行时计算流程图就固定下来不再改变。这种设定对于常用固定模型的计算机视觉任务来说比较方便，相当于特征提取器，其运行时行为较为确定。而且由于模型设置文件独立于代码，比较方便去共享迁移模型。但是随着模型结构的复杂和深度加深，比如 1001 层的 ResNet 模型，其 protobuf 文件已不可能纯手工编写，需要程序生成。而这种生成还是要借助于高层语言如 Python 接口完成，但既然如此何必还要依赖 protobuf 定义模型，直接用 Python 语言编写就好了，因此这套框架的开发模式越来越无法面对日益复杂的计算需求。

而早在 2007 年另一种基于宣告式编程模式的框架 Theano③已经在发展。Theano 主要由蒙特利尔大学的团队开发。相比于 Caffe，其定义模型的过程就是基于 Python 原生语句编写，将数据流情况和相应的计算过程连接按照顺序写下即可。因此要比 Caffe 更灵活，可以通过组合基础算子构成复杂的模块单元。顺带一提的是，在 Theano 之上发展出了两个高层封装，分别为 Lasagne④和 Keras⑤，便于书写深度学习模型中常见的运算层。其中

---

① https://code.google.com/archive/p/cuda-convnet/
② https://caffe.berkeleyvision.org
③ http://deeplearning.net/software/theano/
④ https://lasagne.readthedocs.io/en/latest/
⑤ https://keras.io

Lasagne 已很可惜不再开发了，而 Keras 则兼容了 TensorFlow[①]后端，并且还成为了 TensorFlow 2.0 的核心接口。二者命运之差别，令人不胜唏嘘。

但是 Theano 本身的问题也很大，首先是其调试困难。由于其将运算流书写完成后，需要明确编译步骤才可以实现真正的运算，因此在发生实际运算过程中的错误时，无法追踪中间结果。而且相关文档和社区支持也很差，导致很多问题无法解决。并且 Theano 并没有完整处理大规模数据的流程和支持，只能应用于小规模数据上的实验。但是其计算图定义和计算分离的思想被后来的 TensorFlow 继承发扬，也算是另一种形式延续下去。

同期，还有诸多其他编程框架。比如，有基于 MATLAB 语言的 MatConvNet[②]框架，基于 Java 语言的 Deeplearning4j[③]框架，基于 Julia 语言的 Mocha.jl[④]框架等等。而我们接下来要介绍 PyTorch 的前身，基于 Lua 语言的 Torch7[⑤]框架也属于这其中的一种。只不过相比于其他框架最终的消亡，Torch7 虽一度不被看好，但却最终脱胎换骨成就了今日 PyTorch 的辉煌。

### 1.2.2　PyTorch 前身：Torch7

2011 年的一篇文章 "Torch7: A Matlab-like Environment for Machine Learning"[13]介绍了 Torch7 框架的情况。从标题中可以看出，Torch7 当时突出的特点是具有如 MATLAB 语言一样的易用性的机器学习框架。Torch7 的高层接口语言使用的是 Lua，是一种轻量级多范式动态编程语言，因此写起来像 MATLAB 一样很直观容易。而且 Torch7 底层又支持 CUDA GPU 计算，并且实现了神经网络常用单元和优化器，因此是一个较为全面完整的面向现代深度学习的计算框架。

之后的 PyTorch 在接口命名上承袭了 Torch7，比如在 Torch7 里面定义一个多层感知器（Multi-Layer Perceptron，MLP）是这样写的：

```
mlp = nn.Sequential()
mlp:add(nn.Linear(100,1000))
mlp:add(nn.Tanh())
mlp:add(nn.Linear(1000,10))
mlp:add(nn.SoftMax())
```

如果有了解 PyTorch 的读者会感觉到非常亲切，因为在 PyTorch 里也是类似的命名写法。因此在当时这套框架其实是兼具易用性和高效性的，但为何后来式微了？我认为主要有两个原因。第一是 Lua 这门语言本身缺乏完善性。虽然抛弃了复杂的数据结构和类机制，但也使得处理一些问题时需要用 Metatable 这种特殊技巧去模拟，反而麻烦。同时

---

① https://www.tensorflow.org
② https://www.vlfeat.org/matconvnet/
③ https://deeplearning4j.org
④ https://github.com/pluskid/Mocha.jl
⑤ http://torch.ch

Lua 本身缺乏完善的包管理系统和数据存储读取机制，因此普及起来难度很大，相当于整体的基础设施都要补充。而第二个原因是 Torch7 本身的设计，它主要按照传统的以神经网络为实现载体的深度学习模型，而这种网络往往是比较简单静态的模块连接关系，而遇到了如动态计算、任意有向无环图（Directed Acyclic Graph，DAG）时便束手无策。因此 Torch7 最终没能成为一个广泛普及的计算框架。

### 1.2.3 Torch7 的重生

时间转眼来到了 2017 年，纽约大学 Yann Lecun 教授之前已加入了 Facebook 并组建了 FAIR，而 Torch7 之前正是由纽约大学团队参与开发，因此其团队中的 Soumith Chintala 在经历使用 Torch7 痛苦之后[1]，决定要重新改造 Torch7。既保留 Torch7 简洁直观的编写风格和强大的后端运算核心，同时借助于 Python 语言的易用性和完善的开源社区，打造了全新可微分编程框架 PyTorch。

起初 PyTorch 的实现上还是按照 Torch7 的思路，即保留后端运算库，将原有的 Lua 接口换为 Python。这样虽然解决了之前所述的一个难点，即 Lua 语言本身的小众性和不完善性，但却依然按照静态模块化编程思路，无法实现动态计算。而且当时 TensorFlow 已经推出，并借助强大的宣传影响力，迅速占据科研界和工业界，此时再推出一个计算框架又有何差异呢？

真正让 PyTorch 获得关注和重视的是其支持动态计算图的自动求导（Auto Differentiation）。这使得 PyTorch 可以实现命令式编程（Imperative Programming），即直接编写数据计算流程，且每次运算过程的中间结果可以直接展示，而不是静态固定。同一时期如 MXNet[2]、DyLib[3]、Chainer[4]等框架也都支持相似功能。这种"通过运行来定义"（define-by-run）的编写模式，无需将计算图的搭建和实际运算过程分离，而且支持运行时动态变化，对于当时多种网络结构单元的实现和科研中的快速迭代需求非常契合。因此虽然在推出之时属于"群雄逐鹿"，但后经 Facebook 团队发展努力，PyTorch 现已成为仅次于 TensorFlow 之后最广泛应用的深度学习框架，并且发展势头强劲，大有超越之势。

## 1.3 PyTorch 与 TensorFlow 对比

本节我们将介绍 PyTorch 与 TensorFlow 之间的异同，以便于更好地比较各自的优劣，供实践中合理的选择。

---

[1] https://www.youtube.com/embed/0eLXNFv6aT8
[2] https://mxnet.apache.org
[3] https://dynet.readthedocs.io/en/latest/
[4] https://chainer.org

## 1.3.1 TensorFlow 简介

在 2015 年由 Google Brain 团队推出的 TensorFlow 深度学习计算框架,可以被视为当年人工智能界的一件盛事,因为这是首次由知名企业的顶尖团队打造的一款面向科学研究和工业开发的全面计算框架。之前我们谈到过,TensorFlow 借鉴了 Theano 的计算图思想,在初代版本中,重点突出静态图定义和运算分离,图编译优化,外加大规模数据处理,多节点多卡分布式计算,业务应用部署框架等多种支持,使其备受科研界和工业界的欢迎。并且 TensorFlow 可以支持使用 Google 内部的高性能计算单元 TPU,在坐拥海量数据的资源优势上,可以说是如虎添翼。从之后的 Google 的研究工作可以看出,TensorFlow 框架的推出极大地拓展了模型的规模和计算量,达到了原来不可想象的地步。比如初代的神经架构搜索工作"Neural Architecture Search with Reinforcement Learning"[12],其中使用了 800 个 K40 GPU 运行 28 天得到搜索结果。而这些正是依赖于 TensorFlow 框架对于这样的大规模集群计算的调度。

但是事情的发展在近一两年发生了微妙的变化。首先,虽然这套静态图定义和运算分离的声明式编程方式,便于部署应用,但是对于科研界中常出现的各种动态随机运算过程并不适用,使得其实现时需要花费很大精力去处理。而且 TensorFlow 发展之初的文档及 API 接口比较混乱。其次,TensorFlow 本身核心只想提供最基础的算子,而更复杂的操作可以由用户组合。但是用户的实现千差万别,效率也不尽相同,使得开源初期对于同一个计算层,出现了多种第三方实现方式,并且都合并了主分支代码,反而造成后来用户的无所适从。因此在 2019 年,TensorFlow 升级至 2.0 版本,简化多种实现并存的情况,并且也引入了动态图计算的 eager 模式,也是为了平衡性能和易用性。

## 1.3.2 动静之争

谈到 PyTorch 和 TensorFlow 最大的差别,肯定会提到动态计算图和静态计算图的差别。事实上,TensorFlow 2.0 引入了可以实现 define-by-run 的动态计算模式,实际也说明了在面向科研领域动态计算优势更大。因为从编写代码角度来说,基于命令式的编程模式更加直观,对于中间运算结果可以实时查看调试。从研发创新角度来说,动态计算可允许设计更多新颖复杂的运算单元。比如,当年我曾在早期实验 TensorFlow 时发现,如果想实现一个基于 RNN 的强化学习中的策略执行者(Actor),那么每一步输出的随机离散决策,将作为下一步输入。这在训练过程中间不仅引入了随机采样步骤,而且 RNN 的每一步输入时都是基于之前结果动态决定的。当时的 TensorFlow 是默认 RNN 所有输入已经事先确定好,因此才不会引起循环图的问题。而在我的应用场景中,这一循环依赖不可避免,我也在痛苦地尝试各种实现方式后不得不放弃。当然静态计算图也有其优势所在。比如适合训练结束后的大规模部署,可以在编译阶段优化中间运算操作,从而减少运算内存等资源开销。而且两种模式也必然会长期并存。比如 MXNet 框架就是主打两种计算模式的混合，可以轻松切换。

### 1.3.3 二者借鉴融合

实际上发展到今天，PyTorch 和 TensorFlow 两个主要框架都在相互借鉴融合，取长补短。比如 TensorFlow 就引入了 eager 模式支持动态计算。而在 2.0 版本中主要利用 Keras 接口，使得搭建网络模型的写法和 PyTorch 很相似，增强了可读性。而 PyTorch 也借鉴吸收了 TensorFlow 里完善的可视化数据分析模块 TensorBoard，并且也推出了类似的预训练模型共享方式 torch.hub。可以说二者都在努力补足自身的短板，争取更多的潜在用户。

### 1.3.4 PyTorch 的优势

二者都具有诸多共通之处，那为何还要在 2020 年搞深度学习选择 PyTorch 框架？这里我想主要说明一下 PyTorch 在科研领域的优势。这里限定的科研还特指以大学研究机构为主的实验室科研环境。如果是在公司研发部门，那可能因为早期的技术栈选择便是 TensorFlow，亦或是其他自研框架，则而后续的研究也较大程度受限于此传统。但是也不妨看看下面所谈到的几条 PyTorch 优势，试着尝尝鲜？

首先 PyTorch 的代码层次清晰，模块划分合理，实现统一一致。比如 PyTorch 主要分为基础矩阵运算部分，实现自动微分的计算图，和面向神经网络的常见计算层和单元。这使得在模型搭建优化流程上较为统一，理解直观方便。

其次 PyTorch 有简单的数据处理，模型保存加载，分布式计算方式。比如为了支持常见计算机视觉领域中的数据增广（Data Augmentation）方法，官方还推出 torchvision 模块包含了多种操作。而模型保存加载则是直观的 save、load_state_dict 方法，并且官方的 torchvision 模块中包含了诸多常见 CNN 的实现及相应预训练权重，为结果复现和后续迁移学习提供了统一公平的比较基准。而分布式计算则是简单地利用 torch.nn.DataParallel 即可实现单机多卡运算，而无需过多考虑多个计算图之间的交互。

还有 PyTorch 有简单安装流程和详细的官方文档。PyTorch 的安装方式较为简单，只需要基于 Python 的包管理系统 pip 安装即可，直接支持 GPU 计算，相关依赖较少。官方文档很清晰。这也源于 PyTorch 本身的精简与克制，没有过早地引入额外功能和支持，基本遇到的问题可以很快在官方文档找到解决方案。

最后是性能上的比较。尽管通常来讲人们认为静态计算图的计算效率经过编译优化后会比动态图更高，但从广泛的实验效果和众多实际体验上来说，PyTorch 的运算速度不差，甚至还会在一些特定任务上超越 TensorFlow[1]。这主要源于 TensorFlow 的实现上涉及

---

[1] https://www.reddit.com/r/MachineLearning/comments/cvcbu6/d_why_is_pytorch_as_fast_as_and_sometimes_faster/

额外开销（overhead），而且当前底层密集矩阵运算都使用 cudnn 的实现，使得占据大部分的运算时间基本相同。因此在相似的性能情形下，PyTorch 所具有的易用性和可读性优点便更加突出。

## 1.4 PyTorch 发展现状

接下来我们简要回顾一下 PyTorch 发展至今的状况，并且准备安装 PyTorch，开启我们的学习之旅。

### 1.4.1 主要版本特点回顾

PyTorch 至笔者写本书时最新发布版本为 1.4，期间主要历经了两大版本的迭代。在 0.4.0 版本之前，PyTorch 还是遵循传统思路，区分 Tensor 和 Variable，即前者只关注于张量计算，后者涉及到自动求导。而在 0.4.0 版本开始，Variable 的概念被弱化，即可微分变成了 Tensor 的一个属性，而不需要单独额外建立 Variable。只有在指定关闭自动求导追踪，才会降级成为原始张量计算。这样的好处在于，更加彻底地拥抱动态计算模式，使得代码编写更加简洁清晰，让 PyTorch 自动求导功能默默的在后台中运行。甚至可以达到熟悉 MATLAB 和 NumPy 等科学运算的人们，以非常少的知识迁移代价，即可掌握 PyTorch 框架。

而其从 1.0 版本开始，PyTorch 丰富了更多对于工业界实际应用的支持，比如推出 JIT 编译工具，提供 C++前端接口，提供全新的多节点分布式运算模块，支持 ONNX 格式便于不同框架模型转换，提供 PyTorch Mobile 可以使模型部署在移动端设备等等。因此可以看出 PyTorch 生态正在完善丰富，弥合科研界与工业界研发和实际部署之间的差距。

### 1.4.2 准备工作

之前铺垫了很多，现在我们终于要开始学习 PyTorch 框架了。首先是安装 PyTorch 框架。在 PyTorch 官网 pytorch.org 上面我们会清晰看到不同使用方式的安装方法。图 1.2 为 PyTorch 官网界面。

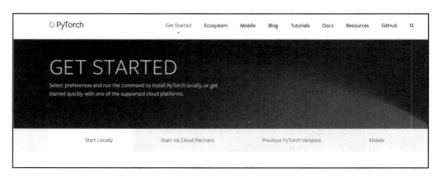

图 1.2

既可以选择在本地服务器个人电脑上安装，也可以使用云服务平台。此处我们选择在本地安装，页面中显示了根据你的操作系统，安装 PyTorch 的版本，是否使用 CUDA GPU 等等条件，自动产生安装命令，如图 1.3 所示。

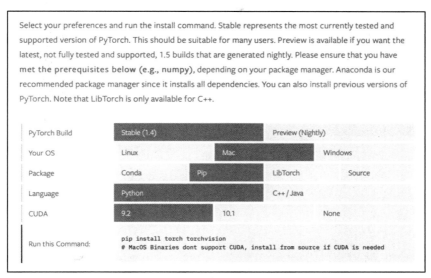

图 1.3

这里我推荐在安装 PyTorch 之前，先安装 Anaconda 这个免费开源的 Python 发现版本，因为其拥有非常方便的包管理部署系统，可以方便切换不同的 Python 版本和安装包版本环境。首先在 www.anaconda.com 官网上下载并安装 Python 3 发行版本。本书中所有代码执行环境只能在 Python 3 下运行，因为 Python 2 已经在 2020 年停止升级开发，而且二者差异不是很多，支持 Python 3 是面向未来更好发展。

在安装好 anaconda 之后，如果可以在命令行中使用如下命令，并反馈正确 anaconda 版本号，说明安装成功：

```
$ conda --version
```

如果提示 conda 命令无法运行，记得将 anaconda 可执行程序的路径~/anaconda3/bin/conda 添加到环境变量中。

然后使用如下 bash 命令行指令创建独立的 Python 包安装环境：

```
$ conda create --name py3.5 python==3.5
```

该指令创建了一个名为 py3.5 的 Python 版本为 3.5 的全新虚拟环境。然后使用如下命令，激活该虚拟环境：

```
$ source activate torch
```

此后所有的包安装管理都在这个单独的环境下。如果退出该环境，则相应安装的包便不存在了。

```
$ source deactivate
```

这样便于使用环境的隔离和管理。在启动 py3.5 虚拟环境后,利用图 1.3 中官网生成的安装命令,根据自身情况与需求,直接安装 torch 和 torchvision 包。

```
$ pip install torch torchvision
```

如果之后在启动 Python 后,可以执行以下语句,并返回了正确的 PyTorch 版本,则说明安装成功:

```
>>> import torch
>>> torch.__version__
```

更多详细的安装步骤和疑问解答可以在 https://pytorch.org/get-started/locally 官网说明中找到。

# 第 2 章 PyTorch 基础计算

本章中我们将介绍 PyTorch 作为深度学习框架的核心：张量计算与自动微分。这两部分也是深度学习的依赖基础。在本章中，我们将会介绍 PyTorch 多种张量运算操作及技巧，以及 PyTorch 自动微分机制。最后我们将会在应用实战中从头搭建简单的神经网络。

## 2.1 PyTorch 核心基础概念：张量 Tensor

作为一个科学运算框架，PyTorch 有着完整的支持张量运算的核心。事实上，如果读者接触过 NumPy 和 MATLAB 等编程库和平台的使用，这一部分知识可以很容易掌握，因为 PyTorch 的许多张量计算操作接口就是仿照 NumPy 的函数用法，便于大家接受理解。但是由于 PyTorch 支持在 GPU 上实现张量运算，因此可以替代 NumPy 极大加速运算过程。

### 2.1.1 Tensor 基本介绍

Tensor 可以理解成为多维数组，高维矩阵，或是 NumPy 中的 ndarray。由于结构化地组织起了大量数据，可以方便进行多种数据操作和数学运算。在 PyTorch 中，初始化一个 Tensor 可以有多种方式，可以从 PyTorch 的列表开始初始化。

```
>>> import torch
>>> torch.tensor([[1., -1.], [1., -1.]])
tensor([[ 1, -1],
        [ 1, -1]])
```

也可以从 NumPy 的 array 转化而来。

```
>>> import numpy as np
>>> torch.tensor(np.array([[1, -1], [1, -1]]))
tensor([[ 1, -1],
        [ 1, -1]])
```

这里 Tensor 的类型是依赖于所初始化列表或者 NumPy 的矩阵类型的。

```
>>> a = torch.tensor([[1, -1], [1, -1]])
>>> a.dtype
torch.int64
>>> b = torch.tensor([[1.0, -1.0], [1.0, -1.0]])
>>> b.dtype
torch.float32
```

而 Tensor 可以通过.int()，.float()，.long()等方法进行相互转化。PyTorch 的 Tensor 的所有数据类型如表 2.1 所示。

表 2.1 PyTorch 中 Tensor 类型总结

| 数据类型 | dtype | CPU 张量表示 |
| --- | --- | --- |
| 32bit 浮点数 | torch.float32 or torch.float | torch.FloatTensor |
| 64bit 浮点数 | torch.float64 或 torch.double | torch.DoubleTensor |
| 16bit 浮点数 | torch.float16 或 torch.half | torch.HalfTensor |
| 8bit 无符号整数 | torch.uint8 | torch.ByteTensor |
| 8bit 有符号整数 | torch.int8 | torch.CharTensor |
| 16bit 有符号整数 | torch.int16 或 torch.short | torch.ShortTensor |
| 32bit 有符号整数 | torch.int32 或 torch.int | torch.IntTensor |
| 64bit 有符号整数 | torch.int64 或 torch.long | torch.LongTensor |
| 布尔型 | torch.bool | torch.BoolTensor |

除了以上初始化张量方式，一些特殊张量可以利用 PyTorch 中的方法快速初始化，如构造全零矩阵。

```
>>> torch.zeros([2, 4], dtype=torch.int32)
tensor([[0, 0, 0, 0],
        [0, 0, 0, 0]], dtype=torch.int32)
```

此处用到了 torch.zeros 方法，并且指定了数据类型为 32bit 整数型。更多的方法使用如下：

```
>>> torch.ones([2, 4])          # 生成全为 1 矩阵
tensor([[1., 1., 1., 1.],
        [1., 1., 1., 1.]])

>>> torch.eye(4)                # 生成对角矩阵
tensor([[1., 0., 0., 0.],
        [0., 1., 0., 0.],
        [0., 0., 1., 0.],
        [0., 0., 0., 1.]])

>>> torch.arange(start=0, end=1, step=0.2)      # 生成[start, end)等距向量
tensor([0.0000, 0.2000, 0.4000, 0.6000, 0.8000])

>>> torch.linspace(start=0, end=1, steps=5)     # 生成[start, end]等距向量
tensor([0.0000, 0.2500, 0.5000, 0.7500, 1.0000])

>>> torch.empty(2, 3)           # 生成未初始化的指定形状矩阵
tensor([[ 0.0000e+00, -2.0000e+00,  3.9368e-32],
        [-1.0845e-19, -3.9674e-07,  4.5783e-41]])
```

```
>>> torch.full(size=[2, 3], fill_value=0.5)    # 利用指定值填充矩阵
tensor([[0.5000, 0.5000, 0.5000],
        [0.5000, 0.5000, 0.5000]])

>>> torch.rand(2, 5)                            # 生成[0, 1)均匀随机采样矩阵
tensor([[0.8707, 0.7948, 0.2776, 0.1043, 0.8918],
        [0.8738, 0.0326, 0.4317, 0.7206, 0.3677]])

>>> torch.randint(low=0, high=10, size=(3, 4))  # [low, high)随机采样整数矩阵
tensor([[1, 1, 8, 9],
        [6, 6, 3, 8],
        [3, 9, 4, 0]])

>>> torch.randn(3, 4)         # 生成标准正态分布采样矩阵
tensor([[-0.1726, -1.1750, -2.4244,  0.0690],
        [ 0.2211,  1.0726, -0.7768,  0.2774],
        [ 0.4949, -1.9298, -0.7551, -0.9732]])
```

PyTorch 中还存在一些对于以上方法名字后边加上 _like 的方法，如 torch.zeros_like、torch.ones_like，其用法如下：

```
>>> a = torch.rand(2, 5)
>>> torch.zeros_like(a)        # 生成和 a 形状一样的全零矩阵
tensor([[0., 0., 0., 0., 0.],
        [0., 0., 0., 0., 0.]])
>>> torch.zeros(a.size())      # 另一种写法
tensor([[0., 0., 0., 0., 0.],
        [0., 0., 0., 0., 0.]])

>>> torch.ones_like(a)         # 生成和 a 形状一样的全一矩阵
tensor([[1., 1., 1., 1., 1.],
        [1., 1., 1., 1., 1.]])
>>> torch.ones(a.size())       # 另一种写法
tensor([[1., 1., 1., 1., 1.],
        [1., 1., 1., 1., 1.]])
>>> a.size()                   # 获取 a 形状的方法
torch.Size([2, 5])
```

注意此处获取张量形状方法为.size()，在后来的版本更新中，加入了.shape 属性（而不是方法），为了与 NumPy 写法保持一致。

PyTorch 中还有一些对于 Tensor 的操作是 Tensor 类的成员方法，并且名称是以 "_" 为结束的。比如：

```
>>> a = torch.randn(3, 4)
>>> a
tensor([[ 0.8985,  0.6567,  0.1452,  1.4166],
        [ 0.3725,  0.6818,  0.3016,  1.1799],
        [-0.3870, -0.5099, -0.1827, -0.2007]])
>>> b = a.zero_()              # 将 a 矩阵原地(inplace)清零，并返回结果为 b
```

```
>>> a                                    # a 矩阵全为 0
tensor([[0., 0., 0., 0.],
        [0., 0., 0., 0.],
        [0., 0., 0., 0.]])
>>> b                                    # b 矩阵也全为 0
tensor([[0., 0., 0., 0.],
        [0., 0., 0., 0.],
        [0., 0., 0., 0.]])
```

这些方法都表示原地操作,即对原张量的元素进行更改。而此时两个矩阵其实共享同一块底层数据存储部分,在 PyTorch 中称为 torch.Storage。

```
>>> a[1][2] = 1
>>> a                                    # a[1][2]元素改变为 1
tensor([[0., 0., 0., 0.],
        [0., 0., 1., 0.],
        [0., 0., 0., 0.]])
>>> b                                    # b 中相应元素也发生了改变
tensor([[0., 0., 0., 0.],
        [0., 0., 1., 0.],
        [0., 0., 0., 0.]])
```

这种机制说明了 Tensor 实际上类似于指向 Storage 的指针。表面上看似两个不同 Tensor,但实际共用一块数据段。这种做法某些时候会引起意外的错误,需要注意。因此类似于深拷贝和浅拷贝的两种不同机制,在 PyTorch 中想要深拷贝一个 Tensor 需要用到.clone()方法。

```
>>> a = torch.randn(3, 4)
>>> b = a.clone()
>>> a
tensor([[ 1.2740, -1.5469, -0.1096, -1.9925],
        [-0.1473,  0.3671, -0.3919, -0.7088],
        [ 0.0469,  0.3644,  1.6394,  2.1083]])
>>> b                                    # b 此时与 a 相同
tensor([[ 1.2740, -1.5469, -0.1096, -1.9925],
        [-0.1473,  0.3671, -0.3919, -0.7088],
        [ 0.0469,  0.3644,  1.6394,  2.1083]])
>>> a.zero_()                            # a 原地清零
tensor([[0., 0., 0., 0.],
        [0., 0., 0., 0.],
        [0., 0., 0., 0.]])
>>> a
tensor([[0., 0., 0., 0.],
        [0., 0., 0., 0.],
        [0., 0., 0., 0.]])
>>> b                                    # 但 b 保持不变,二者无关联
tensor([[ 1.2740, -1.5469, -0.1096, -1.9925],
        [-0.1473,  0.3671, -0.3919, -0.7088],
        [ 0.0469,  0.3644,  1.6394,  2.1083]])
```

## 2.1.2 Tensor 数学运算操作

针对 Tensor，PyTorch 实现了绝大部分的数学运算操作，如果熟悉 NumPy 和 MATLAB 的相关科学计算库接口，可以容易找到对应的 PyTorch 实现。PyTorch 官方文档中将数学运算操作分为六大类，分别为逐点运算子（Pointwise Ops）、规约运算子（Reduction Ops）、比较运算子（Comparison Ops）、谱运算子（Spectral Ops）、其他运算（Other Operations）、BLAS 和 LAPACK 运算。由于所涵盖的运算操作太多，我们只挑常见有代表性的运算操作进行讲解。

对于逐点运算子，其针对 Tensor 中每一元素进行相同运算操作，比如：

```
>>> a = torch.randn(2, 3)
>>> a
tensor([[-0.5239, -0.0788,  1.0364],
        [-0.3985,  0.5109,  2.0730]])
>>> torch.abs(a)                      # 求绝对值操作
tensor([[0.5239, 0.0788, 1.0364],
        [0.3985, 0.5109, 2.0730]])
>>> torch.exp(a)                      # 取 e 指数操作
tensor([[0.5922, 0.9243, 2.8189],
        [0.6713, 1.6668, 7.9484]])
>>> torch.sigmoid(a)                  # sigmoid 函数值
tensor([[0.3719, 0.4803, 0.7382],
        [0.4017, 0.6250, 0.8883]])
>>> torch.clamp(a, min=0, max=1)      # 截断元素值至[0, 1]之间
tensor([[0.0000, 0.0000, 1.0000],
        [0.0000, 0.5109, 1.0000]])
```

而针对以上操作，也有类似于之前以"_"结尾的原地操作，每一次操作将改变原 Tensor 元素值。

```
>>> a
tensor([[-0.5239, -0.0788,  1.0364],
        [-0.3985,  0.5109,  2.0730]])
>>> b = a.abs_()
>>> a
tensor([[0.5239, 0.0788, 1.0364],
        [0.3985, 0.5109, 2.0730]])
>>> c = a.exp_()
>>> a
tensor([[1.6887, 1.0819, 2.8189],
        [1.4896, 1.6668, 7.9484]])
>>> d = a.clamp_(min=1.5, max=2)
>>> a
tensor([[1.6887, 1.5000, 2.0000],
        [1.5000, 1.6668, 2.0000]])
```

以上针对的是一元操作符，而对于二元操作符，PyTorch 中可以使用标准方法的写法，也可以用运算符重载方式简写。注意此处矩阵乘法和普通逐元素乘法的区别。更多关于

矩阵张量乘法运算在后面介绍。

```
>>> a = torch.tensor([[1, 2], [3, 4]])
>>> b = torch.tensor([[5, 6], [7, 8]])
>>> torch.add(a, b)
tensor([[ 6,  8],
        [10, 12]])
>>> a + b
tensor([[ 6,  8],
        [10, 12]])
>>> torch.mul(a, b)              # 逐元素乘法
tensor([[ 5, 12],
        [21, 32]])
>>> a * b
tensor([[ 5, 12],
        [21, 32]])
>>> torch.mm(a, b)               # 矩阵乘法(matrix multiplication)
tensor([[19, 22],
        [43, 50]])
```

更多逐点运算子方法总结如表 2.2 所示。

表 2.2　PyTorch 中逐点运算子方法

| abs | ceil | expm1 | neg | sin |
|---|---|---|---|---|
| acos | clamp | floor | pow | sinh |
| add | cos | fmod | reciprocal | sqrt |
| addcdiv | cosh | frac | remainder | tan |
| addcmul | div | lerp | round | tanh |
| asin | erf | log | rsqrt | trunc |
| atan | erfinv | log1p | sigmoid | |
| atan2 | exp | mul | sign | |

对于规约运算子，指的是对于某一张量进行操作得出某种总结值，如常见包括求最大值指标、求平均值、求方差、求和等等。这里引入了 dim 的概念，即按不同维度进行规约操作。

```
>>> a = torch.randn(2, 3)
>>> a
tensor([[-1.1562,  0.0258,  0.6656],
        [ 1.3922, -0.7169, -0.0335]])
>>> torch.argmax(a)                     # 求全局最大值指标
tensor(3)
>>> torch.argmax(a, dim=1)              # 按照 dim=1 维度，即按行求最大值指标
tensor([2, 0])
>>> torch.sum(a)                        # 求和
tensor(0.1770)
>>> torch.sum(a, dim=0)                 # 按照 dim=0 维度，即按列求和
tensor([ 0.2360, -0.6911,  0.6321])
```

更多规约运算子方法总结如表 2.3 所示。

表 2.3　PyTorch 中规约运算子方法

| argmax | mean | prod | unique |
| argmin | median | std | unique_consecutive |
| dist | mode | std_mean | var |
| logsumexp | norm | sum | var_mean |

对于比较运算子，是指常见的多种张量之间的比较操作，如比较大小排序等。这里特别推荐 PyTorch 独有的 torch.topk 算子，它可以按照给定维度返回张量中最大的 k 个值即对应的指标，在如 ImageNet 数据集中计算 top-5 准确率时非常方便。

```
>>> a = torch.randn(3, 4)
>>> a
tensor([[ 2.5967, -1.6016,  0.7444,  0.2047],
        [ 2.0413, -1.3723, -0.0883,  0.3572],
        [ 0.6732, -2.1633, -0.9910, -2.4277]])
>>> torch.ge(a, 1)                          # 大于等于 (greater equal)
tensor([[1, 0, 0, 0],                       # 返回是uint8型的index
        [1, 0, 0, 0],
        [0, 0, 0, 0]], dtype=torch.uint8)
>>> a >= 1                                  # 运算符重载，与torch.ge相同
tensor([[1, 0, 0, 0],
        [1, 0, 0, 0],
        [0, 0, 0, 0]], dtype=torch.uint8)
>>> a >= torch.zeros(3, 4)                  # 可以允许两矩阵逐元素比较
tensor([[1, 0, 1, 1],
        [1, 0, 0, 1],
        [1, 0, 0, 0]], dtype=torch.uint8)
>>> torch.max(a, dim=1)                     # 返回最大值和对应的指标
(tensor([2.5967, 2.0413, 0.6732]),          # 注意：与numpy中用法不同
tensor([0, 0, 0]))
>>> torch.topk(a, dim=1, k=2)               # 返回按照dim=1 最大 k=2 的值和指标
(tensor([[ 2.5967,  0.7444],
        [ 2.0413,  0.3572],
        [ 0.6732, -0.9910]]),
tensor([[0, 2],
        [0, 3],
        [0, 2]]))
```

更多比较运算子方法总结如表 2.4 所示。

表 2.4　PyTorch 中比较运算子方法

| eq (==) | gt (>) | lt (<) | ne (!=) |
| equal | kthvalue | max | sort |
| ge (>=) | le (<=) | min | topk |

对于谱运算子，指的是一些涉及信号处理傅里叶变换的操作，如快速傅里叶变换（torch.fft）和其逆变换（torch.ifft）、Bartlett 窗函数、Blackman 窗函数、Hamming 窗函数等等。相关内容的理解需要信号处理知识背景，在此不过多讲解。有需要使用相关函数的读者可以参考官方文档[①]。

对于其他运算操作，此处涵盖了多种运算操作，这里选取几个具有代表性常用的运算进行讲解。torch.cumprod、torch.cumsum 等为累积（cumulative）操作，分别代表累积取乘积，取和，与 NumPy 中用法相似。

```
>>> a = torch.tensor([[1, 2, 3], [4, 5, 6]])
>>> torch.cumsum(a, dim=1)          # 按照 dim=1 进行累加[a1, a2, a3]-->
tensor([[ 1,  3,  6],               # [a1, a1+a2, a1+a2+a3]
        [ 4,  9, 15]])
>>> torch.cumprod(a, dim=1)         # 按照 dim=1 进行累乘[a1, a2, a3]-->
tensor([[  1,   2,   6],            # [a1, a1*a2, a1*a2*a3]
        [  4,  20, 120]])
```

还有一部分运算实现矩阵算法，如对角化操作 torch.diag，取下三角矩阵 torch.tril，求矩阵迹 torch.trace 等等。

```
>>> a = torch.tensor([1, 2, 3])
>>> torch.diag(a)                   # 当 a 为向量时，torch.diag 为对角化操作
tensor([[1, 0, 0],
        [0, 2, 0],
        [0, 0, 3]])
>>> b = torch.randn(3, 3)
>>> b
tensor([[-1.8312, -0.0514, -1.9846],
        [ 0.2903,  0.1096,  1.3064],
        [-1.0951, -1.9383, -0.1260]])
>>> torch.diag(b)                   # 当 b 为方矩阵时，torch.diag 为取对角操作
tensor([-1.8312,  0.1096, -0.1260])
>>> torch.tril(b)                   # 取下三角矩阵
tensor([[-1.8312,  0.0000,  0.0000],
        [ 0.2903,  0.1096,  0.0000],
        [-1.0951, -1.9383, -0.1260]])
>>> torch.trace(b)                  # 求迹，和 torch.sum(torch.diag(b))相同
tensor(-1.8477)
```

这里还有一个特殊的运算子 torch.einsum，名为爱因斯坦求和约定（Einstein summation convention）。其原本发源于爱因斯坦在研究广义相对论中烦琐的张量求和计算时，引入的一种简化记号。这种求和约定可以涵盖多种向量、矩阵、张量之间的求和累积操作。简单理解，就是当多个变量中出现相同指标时，代表按此指标进行求和，相当于该指标所代表维度被合并了。这里展示几个例子：

---

① https://pytorch.org/docs/stable/torch.html#spectral-ops

```
>>> a = torch.randn(2, 3)
>>> b = torch.randn(3, 4)
>>> torch.einsum('ij,jk->ik', a, b)          # 代表按照 j 维度进行求和
tensor([[-2.6752,  0.2372,  1.4162,  0.6612],    # 相当于 c_ik=∑_j a_ij*b_jk
        [-0.0152,  1.2279, -1.0661,  2.0650]])
>>> torch.mm(a, b)                            # 等价于 a 和 b 矩阵相乘
tensor([[-2.6752,  0.2372,  1.4162,  0.6612],
        [-0.0152,  1.2279, -1.0661,  2.0650]])
>>> torch.einsum('ij->i', a)                  # 代表按照 j 维度求和
tensor([-1.4899,  1.1475])
>>> torch.sum(a, dim=1)                       # 等价于按照矩阵行求和
tensor([-1.4899,  1.1475])
```

但是要注意的是，虽然 einsum 可以涵盖多种求和操作，但是由于其普适性，导致并未对矩阵乘法进行特殊优化，因此在性能上会差于特定的矩阵乘法操作。einsum 更适合处理较为烦琐的情况，是一种快速实现验证想法的工具。而之后可以在保证结果准确性的基础上再特别优化。感兴趣的读者可以进一步阅读相关介绍文章[1][2]进行学习。更多其他运算操作方法总结如表 2.5 所示。

表 2.5 PyTorch 中其他运算操作方法

| bincount | cummax | diagflat | histc | trace |
|---|---|---|---|---|
| broadcast_tensors | cummin | diagonal | meshgrid | tril |
| cartesian_prod | cumprod | einsum | renorm | tril_indices |
| cdist | cumsum | flatten | repeat_interleave | triu |
| combinations | diag | flip | roll | triu_indices |
| cross | diag_embed | rot90 | tensordot | |

对于 BLAS 和 LAPACK 运算，指的是 PyTorch 实现了基础线性代数程序集（Basic Linear Algeria Subprograms）和线性代数包（Linear Algeria PACKage）中定义的用于数值计算的程序接口，主要实现各种矩阵张量计算，如矩阵乘法、求特征值、求逆、SVD 分解等等。比如最常见的矩阵乘法，如之前所述，可以使用 torch.mm。而有些时候需要实现一种批量矩阵乘法（Batch Matrix Multiplication），如现在流行的 Transformer 等基于 Attention 的语言模型中就大量存在。其描述的是两个三维矩阵 $A$（维度 $b \times n \times m$）和矩阵 $B$（维度 $b \times m \times p$）按照一一对应的 $b$ 组矩阵相乘得到结果矩阵 $C$（维度 $b \times n \times p$）。直接使用这个算子可以免去利用循环实现方式导致的性能下降。更多 BLAS 和 LAPACK 运算方法总结如表 2.6 所示。

---

[1] https://obilaniu6266h16.wordpress.com/2016/02/04/einstein-summation-in-numpy
[2] https://rockt.github.io/2018/04/30/einsum

表 2.6　PyTorch 中 BLAS 和 LAPACK 运算方法

| addbmm | chain_matmul | geqrf | lstsq | matrix_rank | qr |
| --- | --- | --- | --- | --- | --- |
| addmm | cholesky | ger | lu | mm | solve |
| addmv | cholesky_inverse | inverse | lu_solve | mv | svd |
| addr | cholesky_solve | det | lu_unpack | orgqr | symeig |
| baddbmm | dot | logdet | matmul | ormqr | trapz |
| bmm | eig | slogdet | matrix_power | pinverse | triangular_solve |

对于这一部分知识，可以按照实际应用中面临的问题，探索式渐进学习，不求全面掌握。以上所介绍的多种数学运算操作的完整的解释说明可参见官方文档[①]。

## 2.1.3　Tensor 索引分片合并变换操作

PyTorch 中另一大部分操作即是针对 Tensor 的各种索引分片合并变换等操作了。这一部分操作的熟练掌握是迈向高效科学计算的基础，可以极大地加速并精简计算流程。因此在学习这一部分知识时，要试着抛弃之前的循环判断分支等编程思路，用向量化（Vectorize）思维去思考。

对于索引分片操作来说，可以指根据某种条件获取特定值索引，或者根据已有索引值对张量进行赋值。比如常见的根据某一索引获取张量中的值。

```
>>> a = torch.randn(3, 4)
>>> a
tensor([[ 0.6160,  1.1247, -0.1403,  0.8744],
        [ 0.6767, -0.4200, -0.1530,  1.5473],
        [-0.0481, -0.8626,  0.7478,  0.7308]])
>>> torch.index_select(a, dim=1,           # 按照列维度，取第 0 列和第 2 列元素
        index=torch.tensor([0, 2]))
tensor([[ 0.6160, -0.1403],
        [ 0.6767, -0.1530],
        [-0.0481,  0.7478]])
>>> a[:, [0, 2]]                           # 简写。: 表示各行元素均取，即按列取
tensor([[ 0.6160, -0.1403],
        [ 0.6767, -0.1530],
        [-0.0481,  0.7478]])
```

这一部分的内容和符号使用方法，和 NumPy 中的分片索引使用方法一致。运用 ":" 符号代表某一维度元素均被使用，可以起到简化写法的目的。

而对于 Tensor 的合并操作，则包含了拼接（concatenate）、堆叠（stack）、分块（chunk）、拆分（split）等操作。用法如下：

```
>>> a = torch.zeros(2, 3)
```

---

① https://pytorch.org/docs/stable/torch.html#math-operations

```
>>> b = torch.zeros(2, 3).fill_(1)
>>> c = torch.zeros(2, 3).fill_(2)
>>> torch.cat([a, b, c], dim=0)           # 按照 dim=0 进行拼接
tensor([[0., 0., 0.],
        [0., 0., 0.],
        [1., 1., 1.],
        [1., 1., 1.],
        [2., 2., 2.],
        [2., 2., 2.]])
>>> torch.cat([a, b, c], dim=0).shape     # 拼接后不增加额外维度，保持二维
torch.Size([6, 3])
>>> torch.stack([a, b, c], dim=0)         # 按照 dim=0 进行堆叠
tensor([[[0., 0., 0.],
         [0., 0., 0.]],

        [[1., 1., 1.],
         [1., 1., 1.]],

        [[2., 2., 2.],
         [2., 2., 2.]]])
>>> torch.stack([a, b, c], dim=0).shape   # 堆叠后增加额外维度，变成三维
torch.Size([3, 2, 3])
>>> d = torch.cat([a, b, c], dim=0)
>>> torch.chunk(d, 3, dim=0)              # 按照行维度等分为 3 块
(tensor([[0., 0., 0.],
         [0., 0., 0.]]),
 tensor([[1., 1., 1.],
         [1., 1., 1.]]),
 tensor([[2., 2., 2.],
         [2., 2., 2.]]))
>>> torch.split(d, [1, 3, 2], dim=0)      # 按照行维度拆分为[1, 3, 2]宽度的 3 块
(tensor([[0., 0., 0.]]),
 tensor([[0., 0., 0.],
         [1., 1., 1.],
         [1., 1., 1.]]),
 tensor([[2., 2., 2.],
         [2., 2., 2.]]))
```

最后的变换操作，主要包括矩阵的转置（transpose）、矩阵维度缩减（squeeze）与扩张（unsqueeze）等。

```
>>> a = torch.randn(3, 4)
>>> torch.transpose(a, dim0=1, dim1=0)    # 转置操作，维度 0 和维度 1 互换
tensor([[ 0.7052, -1.2079,  0.7141],
        [-0.6149,  0.8292, -0.0044],
        [-0.1635, -1.5694, -2.1606],
        [ 0.1123, -0.0877,  0.2606]])
>>> b = torch.unsqueeze(a, dim=2)         # 张量 a 在维度 2 上添加新维度
>>> b.shape
```

```
torch.Size([3, 4, 1])
>>> torch.squeeze(b).shape                # 缩减无效维度
torch.Size([3, 4])
```

更多的索引分片合并变换操作总结如表 2.7 所示。

表 2.7　PyTorch 中索引分片合并变换方法

| cat | masked_select | split | take | where |
|---|---|---|---|---|
| chunk | narrow | squeeze | transpose | |
| gather | nonzero | stack | unbind | |
| index_select | reshape | t | unsqueeze | |

### 2.1.4　Tensor 类成员方法

除了之前介绍的多种运算操作，对于 Tensor 类中还包含了许多类成员方法。首先之前所介绍的多种方法既可以从函数角度使用，也可以利用其类成员方法实现。比如：

```
>>> a = torch.randn(2, 3)
>>> torch.sum(a, dim=1)
tensor([-0.6439, -1.2272])
>>> a.sum(dim=1)                          # 与上结果相同
tensor([-0.6439, -1.2272])
>>> b = torch.randn(3, 4)
>>> torch.mm(a, b)
tensor([[-0.4001, -0.4716, -0.1721, -0.2903],
        [-1.7405, -0.7037,  0.1716, -1.4012]])
>>> a.mm(b)                               # 二元操作，另外一个张量作为参数
tensor([[-0.4001, -0.4716, -0.1721, -0.2903],
        [-1.7405, -0.7037,  0.1716, -1.4012]])
```

而有一部分操作是可以按照 Tensor 的类成员方法调用。这里选取几个常见方法进行说明。

首先是 .contiguous() 用来使得经过多种合并变换后的 Tensor 在 Storage 的布局（layout）上呈现连续的情形，以便于后续调用如 BLAS 等矩阵运算，因为这些运算为了优化要求输入数据结构连续。而判断一个张量是否连续，可以用 .is_contiguous() 来进行判断。

另一个较为常用的方法是 .item()，用来从只包含单个值的张量中获取 PyTorch 数值。比如，在进行深度学习模型训练时，会计算某种损失函数（Loss Function）的值，而为了展示该标量值的变化情况时，直接用 .item() 即可获取纯粹的 PyTorch 数值，而不是 PyTorch 中的 Tensor 类型。

```
>>> loss1 = torch.tensor(3)
>>> loss1.shape                           # loss1为0维张量
torch.Size([])
>>> loss1.item()                          # 直接获取其数值
```

```
3
>>> loss2 = torch.tensor([3])
>>> loss2.shape                 # loss2 为 1 维张量，与 loss1 不同
torch.Size([1])
>>> loss2[0]                    # loss2[0]获取的仍是一个 0 维张量
tensor(3)
>>> loss2.item()                # loss2.item()获取纯粹的 PyTorch 数值
3
```

还有和 NumPy 数据转换的方法.numpy()。用法如下：

```
>>> a = torch.randn(2, 3)
>>> a
tensor([[-1.5114, -0.3589,  0.3604],
        [ 0.9960, -0.0924,  1.0073]])
>>> b = a.numpy()               # 转换为 numpy.ndarray 格式
>>> b
array([[-1.511361 , -0.35891676,  0.3604407 ],
       [ 0.9959687 , -0.09235841,  1.0073117 ]], dtype=float32)
>>> c = torch.from_numpy(b)     # 利用 from_numpy 方法从 numpy 数据创建新张量
>>> c
tensor([[-1.5114, -0.3589,  0.3604],
        [ 0.9960, -0.0924,  1.0073]])
```

还有.scatter()方法，指的是按照指定索引指标赋值给张量，常用在进行多分类预测时，使用独热编码（One-Hot Encoding），即将基于整数的类别标签，转换成 0/1 向量。关于 scatter 更复杂的使用方法详见官方文档说明[①]。

```
>>> a = torch.zeros(3, 4)
>>> indx = torch.tensor([[1], [2], [0]]) # 三个类别标签
>>> indx.shape
torch.Size([3, 1])
>>> a.scatter_(1, indx, 1.0)            # 按照维度 1 以 indx 所示标签赋值 1.0
tensor([[0., 1., 0., 0.],
        [0., 0., 1., 0.],
        [1., 0., 0., 0.]])
```

最后是.view()方法，其作用是将一个张量转换成另外一个形状，但要求二者总元素数目相同。而且新产生的张量与原张量共享一块数据 Storage，因此 view 只是返回一个以不同格式读取数据的指针，这点需要注意。

```
>>> a = torch.randn(2, 3)
>>> b = a.view(6, 1)            # b 转换为 Size([6, 1])张量
>>> a
tensor([[ 2.2656, -0.1006,  1.5337],
        [-1.4098, -0.8110,  0.4256]])
>>> b
tensor([[ 2.2656],
```

---

① https://pytorch.org/docs/stable/tensors.html#torch.Tensor.scatter_

```
          [-0.1006],
          [ 1.5337],
          [-1.4098],
          [-0.8110],
          [ 0.4256]])
>>> a[0, 1] = 10                        # a[0, 1]元素被更改
>>> b                                   # b相应位置处也发生变化
tensor([[ 2.2656],
        [10.0000],
        [ 1.5337],
        [-1.4098],
        [-0.8110],
        [ 0.4256]])
```

完整的 Tensor 类成员方法可参考官方文档[①]。

## 2.1.5 在 GPU 上计算

使得 PyTorch 相比于之前科学计算库的优势在于其可以方便的利用 GPU 计算，而且代码编写上无需过多的修改。相较于 1.0 版本之前的写法，PyTorch 在新版中的使用 GPU 计算代码编写风格更加通用，甚至可以用全局变量控制同一份代码在 CPU 和 GPU 上快速切换。

在使用 GPU 前，需要测试是否可以在 PyTorch 中使用 GPU 计算，可以用 torch.cuda.is_available()进行测试，如果返回为 True 证明可以使用。而将 Tensor 在 CPU 和 GPU 上相互迁移，则使用简单的.to()方法，使用如下：

```
>>> torch.cuda.is_available()           # 判断是否可以使用支持CUDA的GPU
True
>>> a = torch.randn(2, 3)
>>> b = a.to('cuda')                    # 将a张量移动至GPU上得到b张量
>>> b
tensor([[ 0.2741,  0.6745,  0.1671],
        [-1.3602, -1.5548,  0.7900]], device='cuda:0')
>>> b.device                            # 显示b张量所在device信息
device(type='cuda', index=0)
>>> c = torch.randn(3, 4, device=b.device)# 直接在b的GPU上创建张量c
>>> c                                   # c张量的device与b相同
tensor([[ 0.1877, -0.4171,  0.8542,  1.2767],
        [ 1.3242,  0.0239,  1.3210, -0.3635],
        [-0.0986, -0.6623,  1.8193,  1.7112]], device='cuda:0')
>>> torch.mm(a, c)                      # 注意：CPU张量与GPU张量不能运算
Traceback (most recent call last):
  File "<stdin>", line 1, in <module>
RuntimeError: Expected object of backend CPU but got backend CUDA for argument #2 'mat2'
>>> torch.mm(b, c)        # b和c可以相乘
```

---

① https://pytorch.org/docs/stable/tensors.html

```
tensor([[ 0.9282, -0.2089,  1.4292,  0.3908],
        [-2.3922,  0.0070, -1.7785,  0.1806]], device='cuda:0')
>>> d = a.to('cuda:1')   # 将 a 张量移动到另一个 GPU 上，多个 GPU 编号从 0 开始递增
>>> torch.mm(d, c)       # d 张量和 c 张量不在同一个 GPU 上，也不能运算
Traceback (most recent call last):
  File "<stdin>", line 1, in <module>
RuntimeError: arguments are located on different GPUs at /pytorch/aten/
src/THC/generic/THCTensorMathBlas.cu:255
```

以上代码中展示了 device 的使用，注意 CPU 张量和 GPU 张量之间无法直接运算，存在于不同 GPU 上的张量之间也无法运算。可以看出 PyTorch 中使用 GPU 方法是很简洁直观的。我们也将在后续介绍更多 GPU 计算技巧。

## 2.2 PyTorch 可微编程核心：自动微分 Autograd

PyTorch 中实现自动微分的模块为 Autograd，是作为其可微分编程框架的核心。Autograd 实现的是反向自动微分系统（Reverse Automatic Differentiation），即在建立起前向的计算流时，PyTorch 中的 Autograd 引擎自动去建立该计算流对应的反向计算图，以实现输出对于输入求导。而由于 PyTorch 实现的是动态图计算，即 define-by-run 的模式，每一次的前向运算都会现场重新快速构建反向计算图，因此可以支持动态分支，甚至改变每一次计算图的大小，而无需在一开始将所有可能的计算分支都建立在静态图当中。想要了解如何实现反向自动微分的读者，可以参考 MXNet 作者陈天奇在华盛顿大学开设课程的讲义[1]，其中配有丰富的讲解以及颇具挑战性的课后习题[2]，可以尝试构建自己的自动微分系统。

### 2.2.1 PyTorch 自动微分简介

PyTorch 中的 Autograd 模块在 0.4 版本之后经历了重大升级。原来之前的写法是构建两个概念 Tensor 和 Variable。其中 Tensor 是纯计算载体，不可实现微分求导。而只有将 Tensor 包装成为 Variable 之后，才可以加入到计算图的追踪当中。而在 0.4 版本之后，Variable 的概念被弱化，可微分性变成了 Tensor 的一个属性，无需特意用 Variable 包装。这简化了代码编写。虽然 Variable 这个接口还保留，但是官方不建议这种写法。

### 2.2.2 可微分张量

首先我们介绍可微分张量这个概念。在之前章节中对于张量我们更多关注于其上的各种运算操作，而实际上如果将 Tensor 的可微分属性置为 True，则 PyTorch 会在后台中默默追踪（Trace）着其后计算过程中中间结果得到的方式。

---

[1] http://dlsys.cs.washington.edu/pdf/lecture4.pdf
[2] https://github.com/dlsys-course/assignment1

```
>>> a = torch.randn(2, 3, requires_grad=True)   # 显式化指定a可导
>>> a
tensor([[-1.1335, -1.2754,  1.1586],
        [-0.6539,  1.2571, -0.4889]], requires_grad=True)
>>> a.requires_grad                              # requires_grad为属性
True
>>> b = a ** 2                                   # ** 为幂运算符号
>>> b
tensor([[1.2848, 1.6267, 1.3424],
        [0.4276, 1.5803, 0.2390]], grad_fn=<PowBackward0>)
>>> b.grad_fn                                    # b拥有grad_fn成员
<PowBackward0 object at 0x102e05cf8>             # 记录其反向求导的函数
>>> a.requires_grad = False                      # 显示关闭a的可导属性
>>> a
tensor([[-1.1335, -1.2754,  1.1586],
        [-0.6539,  1.2571, -0.4889]])
```

可以看出对于 PyTorch 的 Autograd 计算引擎可以根据 Tensor 是否可导属性，自动加入到计算图中，并构建出其对应的反向求导函数。但要注意并非所有的运算操作都可以有对应的反向求导函数。

```
>>> a = torch.randn(2, 3, requires_grad=True)
>>> ind = torch.argmax(a)
>>> ind                                          # 最大元素指标无反向求导函数
tensor(0)
>>> ind.requires_grad                            # 该变量也不可导
False
```

类似如最大最小元素指标，排序指标等都不可导。而如何开发出这些操作的近似梯度也是当前可微编程研究的创新点。

### 2.2.3 利用自动微分求梯度

上面介绍了可微张量的概念，接下来就是如何利用 PyTorch 的 Autograd 机制求梯度。其写法也是很直观，只需要对于最终的输出张量，调用.backward()方法即可。

```
>>> a = torch.randn(2, 3, requires_grad=True)
>>> loss = a.sum()                               # 模拟最终输出损失值
>>> loss.backward()                              # 调用.backward()，进行反传
>>> a.grad                                       # 查看张量a的梯度
tensor([[1., 1., 1.],
        [1., 1., 1.]])
```

在这里由于 loss 是一个标量（维度为 0），因此调用.backward()方法默认是反向传播（backpropagation）起始输入为 1，也即实现了 $\frac{\partial loss}{\partial a}$ 的计算。而如果 loss 不是一个标量时，调用.backward()方法的反向传播起始输入应该与 loss 变量维度一致。

```
>>> a = torch.randn(2, 3, requires_grad=True)
>>> loss = a.sum(dim=0)
>>> loss.shape                                    # a 不为标量,而是向量
torch.Size([3])
>>> loss.backward(torch.FloatTensor([1, 2, 3]))# 反向传播输入与 loss 形状相同
>>> a.grad
tensor([[1., 2., 3.],
        [1., 2., 3.]])
```

当输出为向量时,输出对于输入的梯度其实是在求雅各比矩阵(Jacobian matrix),即对于向量函数 $y = f(x)$ 时,$\frac{\partial y}{\partial x} = J$ 的表达式为:

$$J = \begin{pmatrix} \frac{\partial y_1}{\partial x_1} & \cdots & \frac{\partial y_1}{\partial x_n} \\ \vdots & \ddots & \vdots \\ \frac{\partial y_m}{\partial x_1} & \cdots & \frac{\partial y_m}{\partial x_n} \end{pmatrix}$$

其中 $m, n$ 分别为输出和输入的维度。而当计算输出对于输入的改变时,实际是计算在给定输出的变化向量 $v$ 时向量与雅各比矩阵的积(vector-Jacobian product)$v^T J$。PyTorch 的 Autograd 机制正是在隐式地计算出这个结果。相比于直接计算出完整而复杂雅各比矩阵,自动反向求导可以更灵活快速地计算向量与雅各比矩阵的积,比如可以处理循环和动态分支。

```
>>> a = torch.randn(2, 3, requires_grad=True)
>>> loss = a.abs().sum()
>>> while loss < 100:                             # 由 loss 值动态控制循环次数
...     loss = loss * 2
...
>>> loss
tensor(113.1367, grad_fn=<MulBackward0>)
>>> loss.backward()
>>> a.grad                                        # 依然可以求出 a 的梯度
tensor([[-32., -32., -32.],
        [ 32.,  32., -32.]])
```

而在 PyTorch 中,也提供了显示关闭计算图中某一部分张量的自动求导的语法操作。一种是利用 torch.no_grad() 环境,将一整段代码包括进来,停止 Autograd 引擎追踪计算图。

```
>>> a = torch.randn(2, 3, requires_grad=True)
>>> with torch.no_grad():
...     loss = a.sum()                            # 即使 a 可导,loss 也关闭了自动求导
...
>>> loss                                          # loss 张量无对应的反向求导函数
tensor(1.9373)
>>> loss.requires_grad                            # loss 无法反向求导
False
```

这样的做法往往用在模型测试阶段。因为此时已无需训练，关闭 Autograd 求导追踪，可以节省在前向计算时暂存的中间结果，对于使用 GPU 计算时可以节省显存，并且节省计算量达到加速目的。

而另一种更加微观的操作是调用 Tensor 类的 .detach() 方法，其含义是显示地让 Autograd 引擎将某一张量从所追踪的计算图中排除出去。

```
>>> a = torch.randn(2, 3, requires_grad=True)
>>> b = a.detach()                    # b与a数值相同，但b不可导
>>> b.requires_grad                   # b的requires_grad属性为False
False
>>> loss = a.sum()
>>> loss.backward()
>>> a.grad                            # 仅对于a有梯度计算
tensor([[1., 1., 1.],
        [1., 1., 1.]])
>>> b.grad                            # b梯度为None，不显示
```

这个方法可以实现一个常用技巧，即实现前向与反向过程不一致的运算操作。这里的不一致指的是反向过程与按照正常的自动微分系统推导出的反向函数不同，或者前向过程本身不可导，而人为定义的某种近似反向过程。例如，这里以独热编码（One-Hot Encoding）函数为例。该函数可表示为 $y = \mathbb{S}(x)$，其中 $x$ 为输入概率向量，而 $y$ 仅在输入概率 $x$ 最大分量处取值为 1，其他为 0。对于这种函数其直接实现并不可导。

```
>>> x = torch.tensor([0.1, 0.2, 0.3, 0.4], requires_grad=True)
>>> def onehot(x):
...     y = torch.zeros(x.size())     # y先初始化与x维度相同的零向量
...     y[x.argmax()] = 1             # x最大分量位置置为1
...     return y
...
>>> y = onehot(x)
>>> y
tensor([0., 0., 0., 1.])
>>> y.requires_grad                   # y并不可导
False
```

但是如果我们想要使该函数可导，需要设置近似的反传过程函数。在这里我们利用论文 "Estimating or Propagating Gradients Through Stochastic Neurons for Conditional Computation"[14]中提出的直通估计器（Straight-Through Estimator）来模拟反传梯度，即我们可以认为独热编码函数在前向过程输出独热编码向量，但进行梯度反传时，直接对输入概率向量求导，也就是针对某个最终损失函数值导数 $\frac{\partial loss}{\partial y} \approx \frac{\partial loss}{\partial x}$。因此可以这样实现"可导的"独热编码函数：

```
>>> x = torch.tensor([0.1, 0.2, 0.3, 0.4], requires_grad=True)
>>> def onehot(x):
...     y = torch.zeros(x.size())
```

```
...        y[x.argmax()] = 1
...     return (y - x).detach() + x        # 实现了前向反向计算过程不一致的效果
...
>>> y = onehot(x)
>>> y                                       # y 已有 grad_fn 说明其可导
tensor([0., 0., 0., 1.], grad_fn=<AddBackward0>)
>>> y.backward(torch.randn(4))
>>> x.grad                                  # x 成功得到反向传播的梯度
tensor([-0.3069,  2.0079, -0.5531, -1.6199])
```

以上代码中关键的修改是(y - x).detach() + x，在前向计算过程中即等于 y-x+x=y，和原来效果一样。但是反向传播过程中，由于(y-x)这个变量已经从 Autograd 自动追踪的计算图中排除出去，因此相当于反传梯度到了这里，进一步传递给 x，故从整体上来看实现了可导函数。这种做法被可以应用在隐层变量为离散类别变量的变分自动编码器中（variational autoencoder with discrete categorical latent variables），论文[15] [16]中都有进一步的论述。

## 2.2.4 Function：自动微分实现基础

以上介绍了 PyTorch 中自动微分的用法，但是这里所谓的"自动"又是如何实现的？本质上来说还是"手动"的，因为 PyTorch 无法根据任意的前向过程推导出其反向回传函数。而 Autograd 之所以能实现自动推导，其实是利用 Function 这个抽象类。之前已经介绍过，在前向运算过程中，所得到的中间结果的各个 Tensor，都记录着其如何得到 Function，而 PyTorch 事先已经实现了这些 Function 中的反传方法，调用 Tensor 上的.backward()方法时，其实就是沿着反向传播路径上的各个 Tensor 逐一调用其 Function 中的 backward 方法，从而形成完整的反向传播过程。

同时 PyTorch 开放了 Function 这个类别接口，从而可以自定义新的算子前向反向过程。这里我们先实现一个标准的线性整流函数（Rectified Linear Unit，ReLU）作为示例如何使用 Function 类。ReLU 的定义为 $y = \max(x, 0)$，而反传时 $\frac{\partial loss}{\partial x} = \frac{\partial loss}{\partial y} \odot \mathbb{I}(x > 0)$，其中 $\odot$ 为逐元素相乘，$\mathbb{I}(x > 0)$ 为示性函数，即产生一个只有在 $x > 0$ 的位置处为 1，否则为 0 的二值掩码（binary mask）。则我们可以实现以下类，并保存在 relu.py 文件中：

```
from torch.autograd import Function
import torch

class ReLU(Function):                          # 继承 Function 类
    @staticmethod
    def forward(ctx, input):                   # 输入参数列表，可以多个输入 input
        output = torch.clamp(input, min=0)     # 计算 max(input, 0)
        ctx.save_for_backward(output)          # ctx 保存中间结果，用以反传过程使用
        return output

    @staticmethod
```

```
    def backward(ctx, grad_output):        # grad_output 个数与 output 个数相同
        output = ctx.saved_tensors[0]       # 从 ctx 获取暂存的中间结果
        return (output > 0).float() * grad_output   # 计算反传梯度
```

从上面的代码可以看出，一个继承了 Function 类的新算子，主要实现两个方法 forward 和 backward。这两个方法的参数列表中都有一个 ctx 变量，为上下文变量（context），即可以记录该函数输入输出以及中间运算结果，以方便 forward 和 backward 之间共享变量，减少重复运算。ctx.save_for_backward 方法可以将张量结果存入 ctx.saved_tensors 这个列表里。而在实现 backward 时要注意与 forward 的输入输出结果个数匹配。即 backward 的输入 grad_output 要与 forward 计算返回结果 output 个数相同，而 backward 的返回结果要与 forward 的输入 input 个数相同，并且要一一对应。对于无需求导或不可求导的变量，反传梯度设为 None。

利用上述 relu.py 文件，我们可以验证所实现的 ReLU 是否正确。这里展示了使用自定义算子时，调用 ReLU 类的静态方法 apply 实现计算过程。

```
>>> import torch
>>> from relu import ReLU
>>> x = torch.randn(2, 3, requires_grad=True)
>>> y = ReLU.apply(x)            # 新算子的用法，调用 ReLU 类静态方法 apply
>>> loss = y.sum()
>>> loss.backward()              # 反向传播，默认输入为 1
>>> x
tensor([[ 0.5550,  0.2192, -1.3878],
        [ 1.1594,  0.2409, -0.6106]], requires_grad=True)
>>> x.grad                       # I(x > 0) 处梯度为 1，符合预期
tensor([[1., 1., 0.],
        [1., 1., 0.]])
```

Function 类的好处就是，可以自然地实现反传过程与前向过程不一致的算子。之前我们介绍了使用 .detach() 方法的技巧，但是对于更复杂的场景来说，利用 Function 类接口是最通用的解决方案。这里我们还以 ReLU 为例。在论文 "Striving for simplicity: The all convolutional net"[17] 中介绍了一种新的反向传播过程，称之为引导反传（guided backpropagation），这种新定义的反传过程用来对深度模型预测结果进行归因可视化。在论文中描述了这种新的反向传播过程，其主要改变是对于 ReLU 激活函数，反传时公式为：

$$\frac{\partial \text{loss}}{\partial x} = \frac{\partial \text{loss}}{\partial y} \odot \mathbb{I}(x > 0) \odot \mathbb{I}(\frac{\partial \text{loss}}{\partial y} > 0)$$

即反传时不仅只有输入大于 0 的才有回传梯度，还要求输入的回传梯度也要大于 0 的部分才可以传递下去。因此这可以通过修改 relu.py 中 ReLU 类的 backward 方法来实现：

```
class GuidedReLU(Function):                 # 继承 Function 类
    @staticmethod
    def forward(ctx, input):                # 与标准 ReLU 相同
        output = torch.clamp(input, min=0)
        ctx.save_for_backward(output)
        return output
```

```
@staticmethod
def backward(ctx, grad_output):
    output = ctx.saved_tensors[0]        # 以下为修改部分
    mask = (output > 0).float() * (grad_output > 0).float()
    return mask * grad_output
```

同样我们可以验证所实现的新激活函数是否正确。

```
>>> from relu import GuidedReLU
>>> import torch
>>> x = torch.randn(2, 3, requires_grad=True)
>>> y = GuidedReLU.apply(x)
>>> grad_y = torch.randn(2, 3)
>>> y.backward(grad_y)
>>> grad_y
tensor([[ 1.4906,  0.9984, -0.3941],
        [-1.5351, -0.7208,  0.6446]])
>>> x
tensor([[ 0.4234,  0.2058, -2.1999],
        [-0.5862, -0.9008, -0.3732]], requires_grad=True)
>>> x.grad                               # 只有 grad_y 和 x 同时
tensor([[1.4906, 0.9984, -0.0000],       # 大于 0 的位置处才有梯度
        [-0.0000, -0.0000, 0.0000]])
```

然后我们利用此激活函数替换标准的 GoogLeNet[18]中的 ReLU，可视化出来新的引导反传过程产生的梯度与标准反传梯度的差别。图 2.1 中展示了在归因解释图片中真实类别时二者的差异，引导反传梯度会更加的关注于图片中的前景物体，减少了背景的噪声。（相关代码链接：https://github.com/yulongwang12/visual-attribution/blob/master/notebooks/saliency_comparison.ipynb）

真实类别：大象　　　标准反传梯度　　　引导反传梯度

图 2.1

## 2.2.5 注意事项

在使用 PyTorch 中的 Autograd 模块时，还有一些注意事项。首先是梯度累积（gradient accumulation）问题。比如有时候一个 Tensor 可能在多个计算分支中出现，每个不同分支都有相应的损失函数，并且都进行了梯度回传过程。PyTorch 在此处的做法，是将多个分支处的回传梯度累积在一起，而并不是后一次的回传梯度值覆盖掉前一次。因此如果想

要分别得到两次不同回传梯度值,需要在进行第二次回传之前,清空第一次的回传梯度,做法是调用该可导张量的.grad.zero_()方法即可。示例如下:

```
>>> a = torch.randn(2, 3, requires_grad=True)
>>> loss1 = a.sum()
>>> loss2 = (a ** 2).sum()
>>> loss1.backward()
>>> a.grad                                      # 第一次梯度
tensor([[1., 1., 1.],
        [1., 1., 1.]])
>>> loss2.backward()
>>> a
tensor([[-0.5852,  0.8065, -1.0451],
        [-1.1602,  1.3185, -2.0497]],
       requires_grad=True)
>>> a.grad                                      # 两次梯度的累积和
tensor([[-0.1705,  2.6130, -1.0902],
        [-1.3205,  3.6371, -3.0993]])
>>> 2 * a + 1                                   # loss1导致回传梯度为1,
tensor([[-0.1705,  2.6130, -1.0902],            # loss2导致回传梯度为2*a,
        [-1.3205,  3.6371, -3.0993]],           # 二者相加等于a.grad
       grad_fn=<AddBackward0>)
>>> a.grad.zero_()                              # 将a梯度清零
tensor([[0., 0., 0.],
        [0., 0., 0.]])
>>> loss2 = (a ** 2).sum()
>>> loss2.backward()
>>> a.grad                                      # 这次为loss2单独回传的梯度
tensor([[-1.1705,  1.6130, -2.0902],
        [-2.3205,  2.6371, -4.0993]])
```

在上面的代码中,第二次单独求 loss2 产生的梯度时,我们并没有利用之前已有的 loss2 变量直接反传,而是重新计算了一遍 loss2。这么做的原因是在默认情况下,构建好的前向运算图只能回传一次,并且将缓存的反向图释放掉了,不可以回传两遍。而如果想要保留以方便在此回传,需要使用 retrain_graph 属性。

```
>>> a = torch.randn(2, 3, requires_grad=True)
>>> loss = (a ** 2).sum()
>>> loss.backward()
>>> loss.backward()                             # 可以看到出现运行时错误
Traceback (most recent call last):
  File "<stdin>", line 1, in <module>
  File".../torch/tensor.py", line 102, in backward
    torch.autograd.backward(self, gradient, retain_graph, create_graph)
  File ".../torch/autograd/__init__.py", line 90, in backward
    allow_unreachable=True) # allow_unreachable flag
RuntimeError: Trying to backward through the graph a second time, but the
buffers have already been freed. Specify retain_graph=True when calling
```

```
backward the first time.
    >>> a = torch.randn(2, 3, requires_grad=True)
    >>> loss = (a ** 2).sum()
    >>> loss.backward(retain_graph=True)      # 第一次反传时令 retain_graph=True
    >>> a.grad
    tensor([[-1.2740,  1.9519,  4.3163],
            [-3.6763,  2.7135, -3.4478]])
    >>> loss.backward()                        # 再次反传依然可以，并且梯度累积
    >>> a.grad
    tensor([[-2.5479,  3.9038,  8.6326],
            [-7.3525,  5.4270, -6.8956]])
```

以上的反向传递过程，最终所需要求导的 Tensor 均为整个前向计算图中的叶子节点（leaf node）。而 PyTorch 中为了节省运算内存，计算图中间过程的 Tensor 均不求其反传梯度。

```
    >>> a = torch.randn(2, 3, requires_grad=True)
    >>> b = a ** 2
    >>> loss = b.sum()
    >>> loss.backward()
    >>> b.grad == None                         # 作为中间变量的 b 在回传时未保留梯度
    True
    >>> a.grad                                 # 只回传到叶子节点，有梯度
    tensor([[ 1.5900, -0.6698,  0.8239],
            [ 0.4223,  3.3275,  2.3167]])
```

而如果在某些情况下，我们需要求得运算过程中间变量的梯度时，就要使用 torch.autograd.grad 方法：

```
    >>> a = torch.randn(2, 3, requires_grad=True)
    >>> b = a ** 2
    >>> loss = b.sum()
    >>> torch.autograd.grad(loss, b, grad_outputs=torch.tensor(2.0))
    (tensor([[2., 2., 2.],                    # 参数列表分别为：输出，待求导变量，输出回传梯度
            [2., 2., 2.]]),)                  # 此为函数，直接返回梯度结果
    >>> a.grad == None                        # a 和 b 的 .grad 成员变量并没有改变
    True
    >>> b.grad == None
    True
```

利用这一方法，PyTorch 还可以实现高阶微分（high-order differentiation）运算，即实现对于梯度视为变量，而进一步求梯度。下面展示如何求一个 Tensor 的三阶导数：

```
    >>> a = torch.randn(2, 3, requires_grad=True)
    >>> loss = (a ** 3).sum()
    >>> for i in range(3):                     # 循环求 3 次
    ...     grad = torch.autograd.grad(
    ...             loss, a, create_graph=True # 需构建图并保持以方便下一次求导
    ...             )[0]                       # 返回 tuple，取第一个元素即梯度
    ...     loss = grad.sum()                  # 本次梯度变量视为下一次求导的 loss
```

```
...
>>> grad                                          # (a^3)''' = 6
tensor([[6., 6., 6.],
        [6., 6., 6.]], grad_fn=<MulBackward0>)
```

最后需要注意的地方是,作为叶子节点的输入 Tensor 在前向运算过程中不要轻易被其他运算过程赋值,因为这会破坏了运算图的追踪,导致回传时无法找到叶子节点。

```
>>> a = torch.randn(2, 3, requires_grad=True)
>>> for i in range(3):
...     a = a ** 2
...
>>> loss = a.sum()
>>> loss.backward()
>>> a.grad == None         # 由于被重复赋值,此时的a视为中间变量,未保留梯度
True
```

而如果想要进行循环操作,可以另建立一个新变量初始化依赖于叶子节点的赋值,这样虽然有循环赋值操作,但是在最终回传梯度时整个计算过程被展开,多次梯度累积在依然被保留的叶子节点上。因此可以将以上代码改为:

```
>>> a = torch.randn(2, 3, requires_grad=True)
>>> b = a                  # 建立新的代理变量
>>> for i in range(3):     # 赋值操作运行在代理变量上
...     b = b**2
...
>>> loss = b.sum()
>>> loss.backward()
>>> a.grad                 # 叶子节点a可以获得梯度
tensor([[-1.6657e+01,  9.8118e+01, -1.6304e-02],
        [ 2.4258e-02, -1.6693e-06,  2.9978e-01]])
```

## 2.3 PyTorch 应用实战一:实现卷积操作

在本节中,我们将综合运用前面介绍的关于 PyTorch 中的张量操作运算和自动微分机制,从头开始自主实现卷积(convolution)操作。并且可以通过和 PyTorch 内置实现好的卷积算子的运算结果相比较,验证实现正确性。此处应用实战,可以锻炼综合运用张量运算操作的能力,并且了解熟悉卷积机制。

### 2.3.1 卷积操作

卷积原属于信号处理中的一种运算,可以提取信号主要成分信息,去除噪声干扰。后引入 CNN 之中,作为从输入中提取特征的基本操作。首先介绍最普通的二维卷积操作。如图 2.2 所示,展示的是一个输入为 4×4,卷积核尺寸(kernel size)为 3×3,输出为 2×2 的卷积结果。原始的卷积定义中需要对于卷积核权重进行逆向旋转操作,但是由于

卷积核本身权重可以学习，这种旋转操作不必要。可以看出卷积操作其实进行滑动窗相乘得到结果。

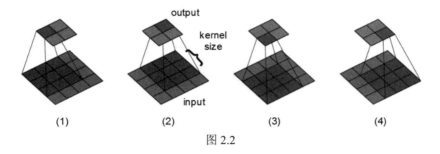

图 2.2

而在此基础上卷积操作引入了补零（zero padding）和跨步（stride）概念。首先补零指的是在输入端外侧填补 0 值以使得卷积输出结果满足某种大小。如图 2.3 所示，展示的是一个输入为 5×5，卷积核为 3×3，补零尺寸（padding size）为 1，输出为 5×5 的卷积结果的前 5 步。可以看出补零就是在原输入的外侧每一边都添加 0 值，目的是使得输出可以达到某种预定形状。

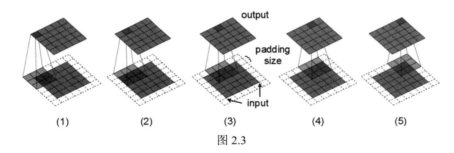

图 2.3

而跨步的概念，则是规定了卷积核在输入上滑动时每次移动到下一步的距离。以上卷积过程都展示的是跨步为 1 的情形。图 2.4 中展示了输入为 5×5，卷积核为 3×3，跨步为 2，无补零的卷积过程，输出为 2×2 情况。可以看到通过调整补零和跨步大小，可以调整不同的输出尺寸。

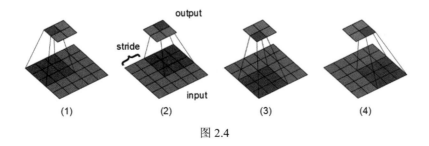

图 2.4

以上的图片均来自于 conv_arithmetic 这个代码库[①]，里面用动画方式直观展示了不同卷积类型的操作过程，还包括更复杂的如转置卷积（transposed convolution）和空洞卷积（dilated convolution）等示意图。鼓励大家进一步探索。

### 2.3.2 利用张量操作实现卷积

虽然我们前面直观展示了卷积的操作，但是实现起来有一些麻烦。这里最主要的是需要处理卷积这个操作中特有的滑动窗操作。因为如果可以有某种方式，能够将每个滑动窗的元素从原始输入张量中提取出来，那么卷积实际上就是这种重整后的张量与卷积核权重之间的矩阵乘法。而在 PyTorch 最新的版本中，开发出来了这样一个操作 torch.unfold，方便实现以上目的。

torch.unfold 可以按照指定维度，以一定的间隔将原始张量进行分片（slicing），然后返回重整后的张量。先以二维矩阵展示其用法：

```
>>> a = torch.arange(16).view(4, 4)
>>> a
tensor([[ 0,  1,  2,  3],
        [ 4,  5,  6,  7],
        [ 8,  9, 10, 11],
        [12, 13, 14, 15]])
>>> b = a.unfold(0, 3, 1)           # 按照行，以每3个元素，跨步为1进行展开
>>> b
tensor([[[ 0,  4,  8],
         [ 1,  5,  9],
         [ 2,  6, 10],
         [ 3,  7, 11]],

        [[ 4,  8, 12],
         [ 5,  9, 13],
         [ 6, 10, 14],
         [ 7, 11, 15]]])
>>> b.shape
torch.Size([2, 4, 3])
>>> c = b.unfold(1, 3, 1)           # 对b按照列以每3个元素跨步为1进行展开
>>> c                               # 注意此时 c 即为 3×3 滑动窗展开结果
tensor([[[[ 0,  1,  2],
          [ 4,  5,  6],
          [ 8,  9, 10]],

         [[ 1,  2,  3],
          [ 5,  6,  7],
          [ 9, 10, 11]]],
```

---

① https://github.com/vdumoulin/conv_arithmetic

```
        [[[ 4,  5,  6],
          [ 8,  9, 10],
          [12, 13, 14]],

         [[ 5,  6,  7],
          [ 9, 10, 11],
          [13, 14, 15]]]])
>>> c.shape
torch.Size([2, 2, 3, 3])
```

从上面的代码可以看出，张量 c 已经实现了按照 3×3 滑动窗大小，跨步为 1 将张量 a 展开成为 4 维张量。因此卷积结果就是让卷积核权重与 2×2 个 3×3 的滑动窗元素相乘并加和得到。

而 CNN 中完整的卷积是对于两个 4 维张量进行操作。其中输入 $x$ 尺寸大小为 $N \times C \times H \times W$，分别代表了批样本量（batch size），输入通道数（channel），输入长和宽。而卷积核 $w$ 尺寸大小为 $D \times C \times K \times K$，分别代表输出通道数、输入通道数和卷积核尺寸（kernel size）。而在每个输出通道上由 $C \times K \times K$ 的单个卷积核与 $C \times H \times W$ 的输入按照通道分别进行二维卷积，再累积在一起形成一个二维的输出。而一个完整的卷积层（convolution layer）则还需在卷积结果上，按照输出通道维度添加偏置值（bias）。更详细的讲解可以参考这里[①]。

以下代码编写在文件 conv.py 中，实现了标准的带补零和跨步设置的 4 维卷积层操作：

```
import torch

def conv2d(x, weight, bias, stride, pad):
    n, c, h_in, w_in = x.shape
    d, c, k, j = weight.shape

    x_pad = torch.zeros(n, c, h_in+2*pad, w_in+2*pad).to(x.device)
    x_pad[:, :, pad:-pad, pad:-pad] = x  # 对输入进行补零操作

    x_pad = x_pad.unfold(2, k, stride)
    x_pad = x_pad.unfold(3, j, stride)   # 按照滑动窗展开
    out = torch.einsum(                  # 按照滑动窗相乘，并将所有输入通道
            'nchwkj,dckj->ndhw',         # 上卷积结果累加
               x_pad, weight
            )
    out = out + bias.view(1, -1, 1, 1)   # 添加偏置值
    return out
```

然后我们可以利用 PyTorch 中已有的卷积操作 torch.nn.functional.conv2d 来验证我们的实现是否正确。

---
① http://cs231n.github.io/convolutional-networks/

```
>>> import torch.nn.functional as F
>>> from conv import conv2d
>>> x = torch.randn(2, 3, 5, 5, requires_grad=True)
>>> w = torch.randn(4, 3, 3, 3, requires_grad=True)
>>> b = torch.randn(4, requires_grad=True)
>>> stride = 2                              # 设定跨步值
>>> pad = 2                                 # 设定补零值
>>> torch_out = F.conv2d(x, w, b, stride, pad)
>>> my_out = conv2d(x, w, b, stride, pad)
>>> print('torch_out == my_out: ',          # 比较自己和 PyTorch 实现的结果
...       torch.allclose(torch_out, my_out, atol=1e-05))
torch_out == my_out: True                   # 二者数值最大误差不超过 1e-5
>>>
>>> grad_out = torch.randn(*torch_out.shape)    # 输出的反传梯度
>>>
>>> grad_x = torch.autograd.grad(           # 求对于输入的梯度
...     torch_out, x, grad_out, retain_graph=True
... )[0]
>>> my_grad_x = torch.autograd.grad(
...     my_out, x, grad_out, retain_graph=True
... )[0]
>>> print('grad_x == my_grad_x: ',          # 比较两种方法反传梯度是否接近
...       torch.allclose(grad_x, my_grad_x, atol=1e-05))
grad_x == my_grad_x: True                   # 二者数值最大误差不超过 1e-5
>>>
>>> grad_w = torch.autograd.grad(           # 求对于权重的梯度
...     torch_out, w, grad_out, retain_graph=True
... )[0]
>>> my_grad_w = torch.autograd.grad(
...     my_out, w, grad_out, retain_graph=True
... )[0]
>>> print('grad_w == my_grad_w: ',
...       torch.allclose(grad_w, my_grad_w, atol=1e-05))
grad_w == my_grad_w: True                   # 二者数值最大误差不超过 1e-5
>>>
>>> grad_b = torch.autograd.grad(torch_out, b, grad_out)[0]
>>> my_grad_b = torch.autograd.grad(my_out, b, grad_out)[0]
>>> print('grad_b == my_grad_b: ',
...       torch.allclose(grad_b, my_grad_b, atol=1e-05))
grad_b == my_grad_b: True                   # 二者对于偏置的反传梯度接近
```

可以看到从头实现的卷积层操作前向计算和反向计算结果均与 PyTorch 内置卷积操作结果相接近，说明了实现的正确性。

## 2.4 PyTorch 应用实战二：实现卷积神经网络进行图像分类

在这一节中，我们将进一步利用上一节所实现的卷积操作，从头开始实现一个简单的卷积神经网络，用以进行 MNIST 数字图像分类任务。首先 MNIST 数字数据集包含了

60 000 张训练数据和 10 000 张测试数据，每张图片是一个 0～9 的手写数字 28×28 的灰度图，任务就是对这些手写数字体进行 10 分类，完整数据集可以在 Yann LeCun 的网站上下载[①]。

这里我们构建一个简单的包含两个隐含层（hidden layer）的卷积神经网络。其网络结构及各层参数如表 2.8 所示，其中 ksize 代表卷积核大小，C_out 为卷积层输出通道数。Flatten 层是将四维张量转化为二维，便于最后计算输出各类概率。

表 2.8 网络结构及各层参数

|  | 参 数 | 输出尺寸 |
| --- | --- | --- |
| Input | - | N×1×28×28 |
| Conv1 | ksize=5, C_out=4, pad=0, stride=2 | N×4×12×12 |
| ReLU | - | N×4×12×12 |
| Conv2 | ksize=3, C_out=8, pad=0, stride=2 | N×8×5×5 |
| ReLU | - | N×8×5×5 |
| Flatten | - | N×200 |
| Linear | num_out=10 | N×10 |

则可以在 cnn.py 文件中如此实现：

```
import torch
import torch.nn.functional as F                    # 计算损失函数
from conv import conv2d
from torchvision import datasets, transforms       # MNIST 数据集加载


def relu(x):                                       # ReLU 激活单元
    return torch.clamp(x, min=0)


def linear(x, weight, bias):                       # Linear 层
    out = torch.matmul(x, weight) + bias.view(1, -1)
    return out


def model(x, params):                              # 所建立的模型
    x = conv2d(x, params[0], params[1], 2, 0)      # stride=2, pad=0
    x = relu(x)
    x = conv2d(x, params[2], params[3], 2, 0)      # stride=2, pad=0
    x = relu(x)
    x = x.view(-1, 200)                            # Flatten 层
    x = linear(x, params[4], params[5])
    return x
```

---

① http://yann.lecun.com/exdb/mnist/

```python
init_std = 0.1
params = [
    torch.randn(4, 1, 5, 5) * init_std,      # Conv1 的 weight 与 bias
    torch.zeros(4),
    torch.randn(8, 4, 3, 3) * init_std,      # Conv2 的 weight 与 bias
    torch.zeros(8),
    torch.randn(200, 10) * init_std,         # Linear 的 weight 与 bias
    torch.zeros(10)
]
for p in params:                             # 设置所有参数可导便于训练
    p.requires_grad = True
```

接下来便是加载数据。这里我们利用 torchvision 中已经提供好的接口，直接加载 MNIST 数据集。有关 torchvision 的使用方法，我们将在第 4 章中进一步介绍。

```python
TRAIN_BATCH_SIZE = 100                       # 训练和测试批样本量
TEST_BATCH_SIZE = 100

train_loader = torch.utils.data.DataLoader(  # 训练 dataloader
    datasets.MNIST(
        './data', train=True, download=True,
        transform=transforms.Compose([       # 输入预处理操作
            transforms.ToTensor(),
            transforms.Normalize((0.1307,), (0.3081,))
        ])
    ),
    batch_size=TRAIN_BATCH_SIZE, shuffle=True) # shuffle 将每一层打乱顺序

test_loader = torch.utils.data.DataLoader(   # 测试 dataloader
    datasets.MNIST(
        './data', train=False,
        transform=transforms.Compose([
            transforms.ToTensor(),
            transforms.Normalize((0.1307,), (0.3081,))
        ])
    ),
    batch_size=TEST_BATCH_SIZE, shuffle=False)
```

接下来我们就可以利用加载好的数据和之前实现的模型进行训练。这里我们使用最简单的随机梯度下降法（stochastic gradient descent，SGD）进行参数优化。即在每次数据前向运算完成后，利用 Autograd 自动求梯度 $f(\theta_t)$，并使用 $\theta_{t+1} = \theta_t - \eta \cdot \nabla f(\theta_t)$ 更新参数，其中 $\eta$ 为学习率（learning rate）。当然实际中我们会运用更复杂更有效的优化器（optimizer），这部分内容将在第 3 章中介绍。

```python
LR = 0.1                                     # 学习率，训练周期，显示训练结果间隔
EPOCH = 100
```

```python
LOG_INTERVAL = 100

for epoch in range(EPOCH):
    for idx, (data, label) in enumerate(train_loader):  # 获得数据和标签
        output = model(data, params)                    # 前向计算
        loss = F.cross_entropy(output, label)           # 计算交叉熵损失函数
        for p in params:
            if p.grad is not None:
                p.grad.zero_()                          # 注意：一定在反传前清除梯度值

        loss.backward()                                 # 反向传播
        for p in params:
            p.data = p.data - LR * p.grad.data  # SGD 更新参数，只能用 .data
                                                # 否则是对于可微变量的赋值
                                                # 破坏前向计算图，导致不可微分

        if idx % LOG_INTERVAL == 0:                     # 打印训练信息
            print('Epoch %03d [%03d/%03d]\tLoss: %.4f' % (
                epoch, idx, len(train_loader), loss.item()
            ))

    correct_num = 0
    total_num = 0
    with torch.no_grad():                               # 测试时关闭自动微分
        for data, label in test_loader:
            output = model(data, params)
            pred = output.max(1)[1]
            correct_num += (pred == label).sum().item()  # 计算累积正确数目
            total_num += len(data)

    acc = correct_num / total_num                       # 计算测试正确率，输出测试结果
    print('...Testing @ Epoch %03d\tAcc: %.4f' % (
        epoch, acc
    ))
```

接下来就可以通过运行 python cnn.py 训练我们从头自己实现的简单 CNN 了。如果一切正常的话，将会输出以下结果：

```
$ python cnn.py
Epoch 000 [000/600]    Loss: 2.3375
Epoch 000 [100/600]    Loss: 0.5527
Epoch 000 [200/600]    Loss: 0.2515
Epoch 000 [300/600]    Loss: 0.3043
Epoch 000 [400/600]    Loss: 0.2315
Epoch 000 [500/600]    Loss: 0.1787
...Testing @ Epoch 000  Acc: 0.9573
Epoch 001 [000/600]    Loss: 0.1019
Epoch 001 [100/600]    Loss: 0.0597
Epoch 001 [200/600]    Loss: 0.1906
```

```
Epoch 001 [300/600]      Loss: 0.1369
Epoch 001 [400/600]      Loss: 0.0844
Epoch 001 [500/600]      Loss: 0.0855
...Testing @ Epoch 001  Acc: 0.9691
Epoch 002 [000/600]      Loss: 0.1364
Epoch 002 [100/600]      Loss: 0.0447
Epoch 002 [200/600]      Loss: 0.1120
Epoch 002 [300/600]      Loss: 0.0850
Epoch 002 [400/600]      Loss: 0.0441
Epoch 002 [500/600]      Loss: 0.0190
...Testing @ Epoch 002  Acc: 0.9750
```

直到训练循环结束。可以看到每一轮测试的准确率在逐步提升。以我的实验结果来看，10 轮训练周期后就可以达到 98%的准确率。大家可以自我探索网络结构和超参数设置，训练达到更高准确率的水平。

# 第 3 章　PyTorch 构建神经网络

在本章中我们将介绍如何利用 PyTorch 高级封装接口快速方便地构建神经网络模型。在上一章，我们展示了一种纯过程式的搭建卷积神经网络的代码。但是可以看出由于其中的卷积层和线性层计算中需要利用到参数变量，所以需要单独维护一个参数列表。而 PyTorch 中将参数和计算封装在一个高级类当中，使得参数初始化，前向过程的计算，利用梯度更新参数等过程变得非常简洁。这也是 PyTorch 一经推出之后广受欢迎的原因。这一章中我们将学习探索这一部分内容，并将在最后的实战应用中展示如何构建复杂结构的模型。

## 3.1　PyTorch 神经网络计算核心：torch.nn

### 3.1.1　nn.Module 概述

PyTorch 中的 nn.Module 类是构建神经网络的基础。Module 类是所有神经网络模块的基类。其他的各种计算层或是你自己实现的层都需要从该类继承。Module 可以被视为一个容器（container），它存储了该层计算所需的各种参数，并且它的内部可以容纳另外一个继承于 Module 的子类形成嵌套。以下的例子显示了如何利用 nn.Module 构建在上一章结尾处我们自己实现的 CNN。这里需要用到 nn.Conv2d、nn.ReLU 和 nn.Linear 各计算单元，实现卷积运算、ReLU 激活函数和线性层操作。我们将在稍后详细介绍。

```
import torch.nn as nn

class SimpleCNN(nn.Module):
    def __init__(self, in_channel=1):
        super(SimpleCNN, self).__init__()    # 注意：需要先调用基类初始化方法
        self.conv1 = nn.Conv2d(
            in_channel, 4, 5, 2, 0      # 输入通道、输出通道、kernel_size
        )                                # stride, padding
        self.relu1 = nn.ReLU()
        self.conv2 = nn.Conv2d(4, 8, 3, 2, 0)
        self.relu2 = nn.ReLU()
        self.linear = nn.Linear(200, 10)    # 输入特征维度，输出特征维度

    def forward(self, x):
        x = self.conv1(x)
        x = self.relu1(x)
```

```
        x = self.conv2(x)
        x = self.relu2(x)
        x = x.view(-1, 200)              # 4 维张量转化为 2 维
        x = self.linear(x)

        return x
```

以上就简单实现了一个 CNN。从中可以看出，一个神经网络在初始化阶段时，其中的各个具体运算层，成为了 Module 继承子类的属性（attribute）。而在具体前向运算时，则在 forward 成员方法中具体写明计算过程即可。此时我们可以测试一下这个 CNN 是否可以正常运算：

```
>>> import torch
>>> model = SimpleCNN()
>>> data = torch.randn(3, 1, 28, 28)     # N x C_in x H x W
>>> output = model(data)
>>> output.shape
torch.Size([3, 10])                      # N x num_classes
```

此处实例化后的 model 是一个可调用（callable）对象，output = model(data)这一句使得model类似于函数接受输入产生输出。这实际上是在调用类的 forward 方法。因此 Module 类很好地统一处理了初始化与运行状态。

此时如果查看 model，可以得到如下打印信息，展示了其内部构成：

```
>>> model
SimpleCNN(
  (conv1): Conv2d(1, 4, kernel_size=(5, 5), stride=(2, 2))
  (relu1): ReLU()
  (conv2): Conv2d(4, 8, kernel_size=(3, 3), stride=(2, 2))
  (relu2): ReLU()
  (linear): Linear(in_features=200, out_features=10, bias=True)
)
```

这里的输出可以看到，在每个计算层前面，有着一个键值（key），是我们之前设置 Module 属性时的名称。实际上，以上的各个计算层的确是以字典（dictionary）形式存储在 Module 的一个隐藏变量里。

```
>>> for k, v in model._modules.items():    # model._modules 为一个字典
...     print(k, v)
...
conv1 Conv2d(1, 4, kernel_size=(5, 5), stride=(2, 2))
relu1 ReLU()
conv2 Conv2d(4, 8, kernel_size=(3, 3), stride=(2, 2))
relu2 ReLU()
linear Linear(in_features=200, out_features=10, bias=True)
>>> type(model._modules)                   # 类型是 OrderedDict,有序字典
<class 'collections.OrderedDict'>          # 即 key 的顺序按照添加到字典里的先后顺序
```

所以在获取 model 中某一层时，既可以直接调用其属性，也可以通过字典按照键值

索引。

以上的展示中感觉 Module 类好像并没有起到额外的作用，感觉就像是有序字典而已。但实际上 Module 还包含了丰富的性质。在建立模型时，我们不仅关注运算过程，还需要利用反向传播的梯度，对网络中的参数进行更新。那如何去获取模型中的参数呢？这就用到了 Module 类中的.parameters()方法：

```
>>> model = SimpleCNN()
>>> model.parameters()
<generator object Module.parameters at 0x102fdc518>
>>> for p in model.parameters():
...     print(p.shape)
...
torch.Size([4, 1, 5, 5])
torch.Size([4])
torch.Size([8, 4, 3, 3])
torch.Size([8])
torch.Size([10, 200])
torch.Size([10])
```

可以看出.parameters()返回的是一个生成器（generator）可以迭代地返回各个可训练参数。那么 Module 又是如何从众多属性成员中识别出其中的可训练参数呢？这里利用到了另外一个特殊类 nn.Parameter。该类的作用可以被视为将一个普通的 Tensor 进行包装标记，使得 Module 类在遍历各成员时，特殊识别出来并加入到自己的隐藏变量._parameters 这个有序字典中。因此我们在实现自己的 Module 子类时，需要将可训练参数用 nn.Parameter 包装起来。

```
>>> class Layer(nn.Module):
...     def __init__(self, param):
...         super(Layer, self).__init__()
...         self.trainable_params = nn.Parameter(torch.randn(2, 3))
...         self.non_trainable_params = param
...
>>> m = Layer(10)
>>> for p in m.parameters():
...     print(p)
...
Parameter containing:                    # 只有 trainable_params 出现
tensor([[-0.2035, -0.5087,  0.9293],
        [ 0.1312, -0.5322,  1.1652]], requires_grad=True)
>>> type(p)                              # p 为 nn.Parameter 类的实例
<class 'torch.nn.parameter.Parameter'>
>>> m.non_trainable_params               # 只是成员属性
10
```

在 2.1.5 节我们介绍了张量在 GPU 上计算的方法。而模型如何迁移到 GPU 上呢？这里 Module 类提供了一个简单的写法就可以实现模型迁移到 GPU 上计算。与 Tensor 的写法一样，都是使用.to()成员方法：

```
>>> model = SimpleCNN()
>>> model = model.to('cuda')                          # 模型迁移到 GPU 上
>>> data = torch.randn(3, 1, 28, 28).to('cuda')       # 数据也同样迁移到 GPU 上
>>> output = model(data)
>>> output.device                                     # 计算结果也在 GPU 上
device(type='cuda', index=0)
```

那么 Module 背后又是什么机制支持可以使用简单的.to()方法即可实现模型参数迁移到 GPU 上？这里细看 Module.to()方法的实现，可以看到调用了 Module 的另一个成员方法.apply()。该方法对 Module 的各个子模块（submodule）迭代式地应用某个操作。比如.to()方法就是迭代式地对每个子模块应用.to()方法，如果是 Tensor 变量则自动迁移到 GPU 上了。类似的 Module 还实现了.float()、.double()、.half()方法，就是逐个将运算参数变成浮点数精度、双浮点精度和半精度。尤其是最后的.half()，不仅可以节省模型占据的显存量，还可以有效加速运算过程。更多利用半精度运算提升速度的实践方法将在后面章节中介绍。

利用.apply()方法还可以实现对模型各层参数的整体初始化。

```
>>> model = SimpleCNN()
>>> def init_weights(m):
...     if m.__class__.__name__ == 'Conv2d':    # 对卷积层权重进行初始化
...         m.weight.data.normal_()
...
>>> model.apply(init_weights)                   # 利用 apply 迭代式初始化
SimpleCNN(
  (conv1): Conv2d(1, 4, kernel_size=(5, 5), stride=(2, 2))
  (relu1): ReLU()
  (conv2): Conv2d(4, 8, kernel_size=(3, 3), stride=(2, 2))
  (relu2): ReLU()
  (linear): Linear(in_features=200, out_features=10, bias=True)
)
```

而在新版的 PyTorch 中，一些常见的初始化算法被总结在了 torch.nn.init 模块中。比如在论文"Delving Deep into Rectifiers: Surpassing Human-Level Performance on ImageNet Classification"[19]中，何恺明等人提出了一种新的初始化算法。如果按此初始化方法，则以上代码可以更改成为：

```
>>> model = SimpleCNN()
>>> def init_weights(m):
...     if m.__class__.__name__ == 'Conv2d':
...         nn.init.kaiming_normal_(m.weight.data)    # 修改了初始化方法
...
>>> model.apply(init_weights)
SimpleCNN(
  (conv1): Conv2d(1, 4, kernel_size=(5, 5), stride=(2, 2))
  (relu1): ReLU()
  (conv2): Conv2d(4, 8, kernel_size=(3, 3), stride=(2, 2))
```

```
    (relu2): ReLU()
    (linear): Linear(in_features=200, out_features=10, bias=True)
)
```

整体更加简洁清晰也准确。

### 3.1.2 结构化构建神经网络

以上我们展示了如何利用继承 Module 类来实现自定义神经网络。但是从写法上来说还是感觉有些麻烦。因为一些常见的模型结构都是单路径直通式结构，即输入逐一经过各计算层得到最终输出，前向计算过程比较单一没有多路分支情况。所以实现的时候可以没必要逐一添加到 Module 类属性，再实现 forward 方法。torch.nn 模块提供了实现这样功能的更多高级容器结构。

首先是 nn.Sequential 类。该类实现的是一个顺序容器（ordered container），即按照添加入其中的模块顺序构建一个单路径网络，顺序执行运算各层操作。像上一节我们构建的 SimpleCNN 模型，可以重新写为：

```
>>> model = nn.Sequential(
...     nn.Conv2d(1, 4, 5, 2, 0),
...     nn.ReLU(),
...     nn.Conv2d(4, 8, 3, 2, 0),
...     nn.ReLU()
... )
>>> model
Sequential(
  (0): Conv2d(1, 4, kernel_size=(5, 5), stride=(2, 2))
  (1): ReLU()
  (2): Conv2d(4, 8, kernel_size=(3, 3), stride=(2, 2))
  (3): ReLU()
)
```

此时的 model 和之前我们利用 Module 所实现的功能完全一样。唯一不足之处是在于，此时获取 model 内部单个计算层有些困难，需要如下这种写法：

```
>>> model._modules['0']
Conv2d(1, 4, kernel_size=(5, 5), stride=(2, 2))
>>> model._modules['1']
ReLU()
```

也就是按照添加顺序的编号作为键值，而无法通过成员属性获取了。为了实现这种效果，可以改写为利用有序字典类（OrderedDict）来搭建模型：

```
>>> from collections import OrderedDict
>>> model = nn.Sequential(OrderedDict([
...     ('conv1', nn.Conv2d(1, 4, 5, 2, 0)),    # key-value pair
...     ('relu1', nn.ReLU()),
...     ('conv2', nn.Conv2d(4, 8, 3, 2, 0)),
```

```
...        ('relu2', nn.ReLU())
... ]))
>>> model
Sequential(
  (conv1): Conv2d(1, 4, kernel_size=(5, 5), stride=(2, 2))
  (relu1): ReLU()
  (conv2): Conv2d(4, 8, kernel_size=(3, 3), stride=(2, 2))
  (relu2): ReLU()
)
>>> model.conv1                              # 以类似成员属性方式获取某一层
Conv2d(1, 4, kernel_size=(5, 5), stride=(2, 2))
>>> model.relu1
ReLU()
```

nn.Sequential 可以看成是一个单独有序的容器，但是仅能建模运算路径单一的网络结构。而有些时候网络中存在着多路分支并行情况。在分支数较少时，还可以利用循环手动写出每一分支计算过程，但是当分支数过多时，这种写法比较麻烦。torch.nn 模块中还提供了另一个容器 nn.ModuleList，既满足了存储多个模块容器，又没有强制规定只能按照单一顺序运算。这里可以将 nn.ModuleList 就理解成为 PyTorch 中的普通列表，可迭代可索引，包含了添加（append）、扩展（extend）、插入（insert）等方法。与普通列表的不同之处在于，它又可以实现 Module 中的一些方法，比如提取子模块中的参数，一键迁移至 GPU 等。下面的例子展示了包含 10 个分支的一个模块，其中每路分支输出通道数递增 1，最终的输出是各路分支输出按照通道维度拼接在一起。

```
class MultiBranch(nn.Module):
    def __init__(self, in_channel=1, n_branch=10):
        super(MultiBranch, self).__init__()
        self.branches = nn.ModuleList([])         # 初始化为空列表
        for k in range(n_branch):
            self.branches.append(                 # 循环添加
                nn.Conv2d(in_channel, k+1, 3, 1, 1)
            )

    def forward(self, x):
        out = []
        for m in self.branches:                   # 循环计算各路分支
            out.append(m(x))

        out = torch.cat(out, dim=1)               # 按照通道维度拼接在一起
        return out
```

下面的代码测试了以上实现的正确性：

```
>>> model = MultiBranch()
>>> model
MultiBranch(
  (branches): ModuleList(
```

```
    (0): Conv2d(1, 1, kernel_size=(3, 3), stride=(1, 1), padding=(1, 1))
    (1): Conv2d(1, 2, kernel_size=(3, 3), stride=(1, 1), padding=(1, 1))
    (2): Conv2d(1, 3, kernel_size=(3, 3), stride=(1, 1), padding=(1, 1))
    (3): Conv2d(1, 4, kernel_size=(3, 3), stride=(1, 1), padding=(1, 1))
    (4): Conv2d(1, 5, kernel_size=(3, 3), stride=(1, 1), padding=(1, 1))
    (5): Conv2d(1, 6, kernel_size=(3, 3), stride=(1, 1), padding=(1, 1))
    (6): Conv2d(1, 7, kernel_size=(3, 3), stride=(1, 1), padding=(1, 1))
    (7): Conv2d(1, 8, kernel_size=(3, 3), stride=(1, 1), padding=(1, 1))
    (8): Conv2d(1, 9, kernel_size=(3, 3), stride=(1, 1), padding=(1, 1))
    (9): Conv2d(1, 10, kernel_size=(3, 3), stride=(1, 1), padding=(1, 1))
  )
)
>>> data = torch.randn(3, 1, 5, 5)
>>> out = model(data)
>>> out.shape                          # 通道维度为 1+2+...+10=55
torch.Size([3, 55, 5, 5])
```

### 3.1.3 经典神经网络层介绍

这一节我们将介绍 torch.nn 模块中更多实现好的经典神经网络计算层。首先最常见的是卷积层，分为一维卷积 nn.Conv1d，二维卷积 nn.Conv2d 和三维卷积 nn.Conv3d。其 api 接口均为(in_channels, out_channels, kernel_size, stride, padding, dilation, groups, bias)，分别代表着输入特征通道数、输出特征通道数、卷积核大小、跨步长度、补零个数、空洞卷积扩张率、分组卷积分组个数和是否添加偏置项。这里重点说一下 groups 这个参数。默认标准卷积是所有输入通道都对每一个输出通道有贡献。而在 AlexNet 早期实现上，为了实现减小显存的目的，引入了分组为 2 的卷积，即将输入分为两组，每一组贡献一半的输出通道。进一步的，在论文 "Aggregated Residual Transformations for Deep Neural Networks"[20]中则探索了 groups 这个变量对于模型最终性能的影响，发现分组数增加可以在相似的计算量下提升模型性能。而当 groups=in_channels 时，即实现了 depthwise convolution。这种卷积现在被广泛运用到多种轻量化模型中，如 Xception[21]、MobileNetV2[22]等。图 3.1 展示了不同 groups 数目对于卷积过程的影响，图片来自于论文[23]。

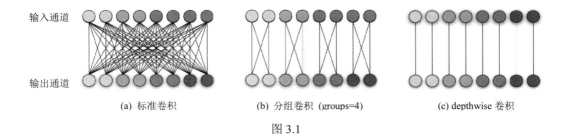

(a) 标准卷积　　　　(b) 分组卷积 (groups=4)　　　　(c) depthwise 卷积

图 3.1

另一个常见运算层为池化层（pooling layer）。与卷积层类似，通过滑动窗方式，对每一个滑动窗内部实施某种规约操作，起到缩减输入特征尺寸的作用。常见的池化层有最大池化层（nn.MaxPool）和平均池化层（nn.AvgPool），分别在滑动窗内部进行最大化操作和求平均操作。这两种层也都有 1D/2D/3D 的版本。其参数列表为(kernel_size, stride, padding, dilation, return_indices, ceil_mode)。其中前面一些概念与卷积相同，唯一区别是最后两个参数。return_indices 控制是否返回输出特征各元素在原输入特征中的索引位置。这个主要是用在反最大池化操作（nn.MaxUnpool）中，即将一个输出特征反向扩展到输入空间大小。ceil_mode 是决定最后输出特征维度的计算时，取整方式是向下还是向上。

接下来是常见的激活单元层（activation layer），这里主要介绍常见的几种。nn.ReLU 是线性整流单元，这个之前已经多次出现。nn.Sigmoid 是实现 sigmoid(x) = 1 / (1 + exp(-x)) 函数输出。nn.ReLU6 类似于 nn.ReLU，只不过限制其最大输出值为 6，保证数值稳定。这种特殊的改进出现在 MobileNetV2[22]结构中被证明有效。nn.LeakyReLU 对 nn.ReLU 的另一种改进，使得其在输入小于 0 的部分不是完全被抑制为 0，而是可以有一定斜率输出。表达式为

$$LeakyReLU(x) = \begin{cases} x, & x \geq 0 \\ nslope \times x, & x < 0 \end{cases}$$

其中 nslope 为控制负输入部分输出的斜率。这种单元常用在生成式模型如 GAN[24]中。更多激活单元类型总结在表 3.1 中。

表 3.1　激活单元类型

| ELU | MultiheadAttention | SELU | Softshrink | Softmin |
|---|---|---|---|---|
| Hardshrink | PReLU | CELU | Softsign | Softmax |
| Hardtanh | ReLU | GELU | Tanh | Softmax2d |
| LeakyReLU | ReLU6 | Sigmoid | Tanhshrink | LogSoftmax |
| LogSigmoid | RReLU | Softplus | Threshold | |

还有线性层 nn.Linear 和 Dropout 层 nn.Dropout。其中前者接受特征向量，通过权重矩阵相乘得到输出，往往出现在模型最后类别输出。而后者则是随机对特征上每个元素置为 0，包含了 1D/2D/3D 版本，可以起到正则化效果，增强模型泛化能力。具体讨论可参见论文[25]。

接下来要重点介绍几种常见的归一化层（normalization layer）。首先是批归一化层 nn.BatchNorm。该层可以有效提升深层模型预测性能。该层计算过程分为训练阶段和测试阶段。在训练阶段，按照通道维度进行如下公式计算：

$$y = \frac{x - E(x)}{\sqrt{Var(x) + \epsilon}} * \gamma + \beta$$

前一部分是将 x 归一化，$\gamma, \beta$ 是进行仿射变换（affine transformation）的参数，是可训练的参数。这里会跟踪记录每一批次的均值方差，并在训练过程中持续更新，但不是

可训练的参数。而到了测试阶段，则直接用训练过程中估计的均值方差进行上式计算。因此这里涉及到两个问题：第一，如何实现一个模块在训练和测试阶段两种不同的运行方式；第二，如何既保留手动更新的均值方差统计量，又不使其成为可训练参数加入到自动求导图中。

为了实现训练和测试过程的不同运算方式，nn.Module 里引入了.training 这个成员变量，用以判断当前状态。同时也提供了.train()和.eval()两个方法，可以递归式地对所有模块设置.training 变量。

```
>>> m = nn.BatchNorm2d(32)
>>> m.train()
BatchNorm2d(32, eps=1e-05, momentum=0.1,
    affine=True, track_running_stats=True)
>>> m.training                              # 通过设置 training 变为 True
True
>>> m.eval()
BatchNorm2d(32, eps=1e-05, momentum=0.1,
    affine=True, track_running_stats=True)
>>> m.training                              # 通过设置 training 变为 False
False
```

要区分这个变量和当前计算图是否可求导是独立的，即可以出现在 torch.no_grad()条件下设置模型.training 变量为 True 的情况，只要此时的行为合理即可。

而为了解决既可以保留部分成员张量作为模型的某种辅助参数，又不使其成为可训练参数加入到计算图中，nn.Module 引入了 buffer 这个概念，通过.register_buffer()方法可以命名某个成员变量。这个成员变量可以通过属性索引，可以计算更新，可以在模型保存时导出成为参数的一部分，但又不是 nn.Parameter 的实例，即不是可导张量。

接下来介绍常见的损失函数层（loss layer）。损失函数层往往出现在模型的输出部分，且在训练阶段启动引导参数学习的作用。因为搭建模型时往往作为独立模块，以便于在测试时避免额外计算。最常见的用于多分类的损失函数为 nn.CrossEntropyLoss，它用来计算输入和标签之间的交叉熵。但要注意这里的输入需要是模型对于各个类别的原始输出值，而非经过如 softmax 函数归一化后的概率值。而标签需要的是指示类别的整数，而非 one-hot 编码形式。

如果是需要计算两个概率分布直接的交叉熵，应该直接使用 $loss = \sum_i -p_i \log q_i$ 公式，其中 $p_i, q_i$ 分别为两个概率分布的各个分量。以下代码验证了两种计算方法的等价性：

```
>>> data = torch.randn(3, 10)
>>> label = torch.randint(0, 10, size=(3,))
>>> label
tensor([1, 8, 2])
>>> loss1 = nn.CrossEntropyLoss()
>>> loss1(data, label)                      # 使用未归一化输出值和整数标签计算交叉熵
tensor(3.2831)
```

```
>>> logprob = nn.LogSoftmax(dim=1)(data)    # 先计算概率的对数值
>>> onehot_label = torch.zeros(3, 10).scatter_(1, label.unsqueeze(1), 
1.0)
>>> onehot_label                             # 再将标签转换成为 onehot 编码
tensor([[0., 1., 0., 0., 0., 0., 0., 0., 0., 0.],
        [0., 0., 0., 0., 0., 0., 0., 0., 1., 0.],
        [0., 0., 1., 0., 0., 0., 0., 0., 0., 0.]])
>>> loss2 = (onehot_label * logprob).sum(dim=1).mean()# 按照每个样本维度平均
>>> loss2                                    # 两种计算 loss 值相同
tensor(3.2831)
```

而在二分类情况下，则直接可以使用二分类交叉熵（Binary Cross Entropy，BCE）损失函数 nn.BCELoss。但这里又要注意的是，nn.BCELoss 接受的输入和标签均需要为二类别概率，与 nn.CrossEntropyLoss 输入格式不同。还有在回归问题中常使用的均值平方差（Mean Squared Error，MSE）损失函数 nn.MSELoss，在物体检测[26]中为防止梯度爆炸而引入的 nn.SmoothL1Loss，语音等时间序列模型问题[27]中使用的 nn.CTCLoss 等等。更多损失函数层类型总结在表 3.2 中。

表 3.2  损失函数层类型

| L1Loss | PoissonNLLLoss | HingeEmbeddingLoss | CosineEmbeddingLoss |
|---|---|---|---|
| MSELoss | KLDivLoss | MultiLabelMarginLoss | MultiMarginLoss |
| CrossEntropyLoss | BCELoss | SmoothL1Loss | TripletMarginLoss |
| CTCLoss | BCEWithLogitsLoss | SoftMarginLoss | |
| NLLLoss | MarginRankingLoss | MultiLabelSoftMarginLoss | |

最后介绍的是循环神经网络层（Recurrent Neural Network Layer）。循环神经网络特点在于输出可以作为输入再次进行运算。因此这里涉及到前向计算图出现循环反馈的问题。而在 PyTorch 中这里每次循环计算的各个可导变量，先是被视为在时间轴上的单个独立变量，而在反传梯度后，将各次求导得到变量累加在一起。因此 RNN 层的反向传播算法比较难实现。在 torch.nn 里面提供两个层次的封装。首先是最核心的循环 Cell 层，即进行单步输入到输出的计算，分别有 nn.RNNCell、nn.LSTMCell、nn.GRUCell。例如，以 nn.LSTMCell（long-short term memory，LSTM）为例，其单步运算过程为：

$$i = \sigma(W_{ii} x + b_{ii} + W_{hi} h + b_{hi})$$
$$f = \sigma(W_{if} x + b_{if} + W_{hf} h + b_{hf})$$
$$g = \tanh(W_{ig} x + b_{ig} + W_{hg} h + b_{hg})$$
$$o = \sigma(W_{io} x + b_{io} + W_{ho} h + b_{ho})$$
$$c' = f * c + i * g$$
$$h' = o * \tanh(c')$$

以上的运算过程被封装在 nn.LSTMCell (input_size, hidden_size, bias)的接口中。更详

细的讲解可以参考该 blog[①]中的可视化过程和论文[28]中推导。另一个层次是包含多个 Cell 层的深度循环网络模型，分别有 nn.RNN、nn.LSTM、nn.GRU，代表了其内部堆叠了多个基本的 Cell 构成的循环神经网络。而且此时将整体处理整个时间序列而不再需要迭代反馈计算结果。

```
>>> seq_len = 5                          # 时间序列长度 L
>>> n_batch = 3                          # 批样本量 N
>>> input_size = 10                      # 输入特征维度 C_i
>>> hidden_size = 7                      # 隐层特征维度 C_h
>>> num_layers = 3                       # 层数
>>> num_directions = 2                   # 双向 LSTM，方向数为 2
>>> model = nn.LSTM(
...     input_size, hidden_size, num_layers,
...     bidirectional=True               # 设置为双向
... )
>>> input = torch.randn(seq_len, n_batch, input_size) # 输入 L x N x C_in
>>> h0 = torch.randn(num_layers * num_directions, n_batch, hidden_size)
>>> c0 = torch.randn(num_layers * num_directions, n_batch, hidden_size)
>>> output, (h, c) = model(input, (h0, c0))
>>> output.shape                         # 输出 L x N x (C_h * 2)
torch.Size([5, 3, 14])
```

### 3.1.4 函数式操作 nn.functional

以上介绍的多种运算模块层，在 torch.nn 模块中也配备了对应的函数式调用接口。比如对于卷积层 nn.Conv2d，如果在某处只需要利用其卷积操作，而不想让 nn.Module 初始化相关参数，可以使用 nn.functional 中的 conv2d 函数来运算。一般 torch.nn.functional 模块简记为 F，以下为示例代码：

```
>>> import torch.nn.functional as F
>>> x = torch.randn(4, 3, 5, 5)          # N x C x H x W
>>> w = torch.randn(7, 3, 3, 3)          # C_o x C x K x K
>>> b = torch.randn(7)
>>> o = F.conv2d(x, w, b, stride=2, padding=1)
>>> o.shape                              # N x C_o x H_o x W_o
torch.Size([4, 7, 3, 3])
```

因此可以将 nn.functional 中的各种运算类别称为复杂高级的 torch 运算操作符。而之前所介绍的多种经典神经网络层模块均有其对应的函数式运算符。

但这里还是要介绍 nn.functional 中特有的一些运算，被统称为视觉函数（visual functions）。主要是这些运算常针对于视觉任务中的上采样插值等。以下我们将介绍一些重点的运算操作。首先是上采样操作 F.upsample，其参数列表为(input, size, scale_factor,

---

① https://colah.github.io/posts/2015-08-Understanding-LSTMs/

mode, align_corners)。其中 input 为输入特征需要为 3D/4D/5D 的形状。size 为指定放缩输出特征的大小，scale_factor 为指定放缩的倍数，因此二者选择其中一个就可以。mode 可以选择在放缩时的插值模式，有'nearest' | 'linear' | 'bilinear' | 'bicubic' | 'trilinear'这几种。但是随着 PyTorch 最新版本更新，官方推荐使用 F.interpolate 函数代替 F.upsample，因为 F.interpolate 可以同时处理放大和缩小特征输出，含义更广泛避免误解。F.interpolate 参数列表与之前介绍的相同。

另外一个函数 F.pad，对输入特征进行填补操作。其参数列表为(input, pad, mode, value)，其中 input 为 N 维输入，pad 为指定某一维度扩展填补的元素个数。这里 pad 的设置稍有复杂。它是指定从最后一个维度并逐一往前的每一维度两侧扩展数量。比如，对最后一个维度扩展，则 pad=(pad_left, pad_right)，这里该维度两侧需要分别指定。如果对最后两个维度扩展，则 pad=(pad_top, pad_bottom, pad_left, pad_right)，以下代码展示了如何使用 F.pad，其中 mode 指定是填补模式，可以选择'constant' | 'reflect' | 'replicate' | 'circular'几种模式，value 指的是填补值。

```
>>> x = torch.randn(4, 3, 5, 5)
>>> x1 = F.pad(x, (1, 2), mode='constant', value=0)
>>> x1.shape                          # 最后一维，两侧共扩展 3 个，故为 5+3=8
torch.Size([4, 3, 5, 8])
>>> x2 = F.pad(x, (1, 2, 3, 4), mode='constant', value=0)
>>> x2.shape                          # 倒数第二维，两侧共扩展 7 个，故为 5+7=12
torch.Size([4, 3, 12, 8])
```

接下来介绍一对函数操作 F.grid_sample 和 F.affine_grid。其中前者是根据给定的一个逐像素偏移矩阵[视频处理领域中该矩阵为光流场（optical flow field）]，将原始图片逐像素扭曲（warp）到目标位置处形成一个新的图片。这种操作可以实现图片的多种变换如旋转、放缩、平移等等。而后者则是根据一个仿射变换矩阵，生成该变换对应的逐像素偏移矩阵或者光流场矩阵。以下例子展示了将一张图片逆时针选择 45°的结果。

```
from PIL import Image
import torch
import torch.nn.functional as F
from torchvision.transforms import ToTensor

def pil_loader(path):                   # 图片读取函数
    with open(path, 'rb') as f:
        with Image.open(f) as img:
            return img.convert('RGB')


img = pil_loader('elephant.png')         # 读取图片
img = ToTensor()(img).unsqueeze(0)       # 转换为 Tensor，形状 NCHW

M = torch.tensor(                        # 仿射变换矩阵
```

```
    [[[0.7071, -0.7071, 0],
     [0.7071,  0.7071, 0]]]
)

grid = F.affine_grid(M, size=(1, 3, 500, 500))     # 生成光流场
warp_img = F.grid_sample(img, grid)                 # 扭曲图片
```

图 3.2 展示了原始图片、光流场和扭曲后的图片。可以看到原始图片每个像素点按照光流场偏移矩阵移动，可以实现图片的旋转。

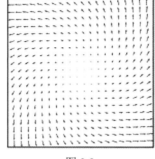

图 3.2

以上两者的配合可以实现 Spatial Transformer Networks[29]。其中仿射变换的矩阵 M 可以建模成为一个神经网络的输出，从而学习得到聚焦于图片中主要物体的仿射变换。这对于处理图片中存在多个尺度不同的物体很有帮助。

## 3.2 PyTorch 优化器

以上所介绍的是如何利用 nn.Module 搭建模型，实现复杂的前向运算。而在训练阶段另一个重要的组成部分便是优化器（optimizer），指的是如何利用自动求导得到的各参数的导数，更新参数。

### 3.2.1 torch.optim 概述

在第 2 章末尾我们利用了最简单的 SGD 优化算法，只是用参数按一定学习率沿着梯度方法更新。而实际上 PyTorch 中实现了更多复杂的优化算法，并被封装在了 torch.optim 模块之中方便调用。torch.optim 一般简记为 optim。

一个优化器首先在初始化时需要接收待训练参数组。这个参数组可以为一个可迭代的 generator 不断返回参数，也可以是一组参数组成的列表，即可以有以下写法：

```
optimizer = optim.SGD(model.parameters(), lr=0.1, momentum=0.9)
optimizer = optim.SGD([params1, params2], lr=0.1)
```

这里 lr、momentum 等参数指定了 SGD 算法中的学习率、动量等参数。而有些时候，针对于不同的参数，我们想要设置不同的学习率等超参数（hyper-parameter）时，可以在

构建优化器时传入字典。

```
optimizer = optim.SGD([
    {'params': model.base.parameters()},
    {'params': model.classifier.parameters(), 'lr': 1e-3}
], lr=1e-2, momentum=0.9)
```

这里可以看到对于 SGD 优化器，有 params、lr 等键值，不同的参数设置不同超参数，而最后的 lr、momentum 则是对所有参数的默认设置超参。

当构建好了优化器之后，便可以进行训练。一般来说训练过程可写作以下方式：

```
model = init_model_function(...)          # 模型构建
optimizer = optim.SomeOptimizer(          # 优化器构建
    model.parameters(), lr=..., mm=...
)

for data, label in train_dataloader:
    optimizer.zero_grad()                 # 前向计算前，清空参数原有梯度

        output = model(data)              # 前向运算，计算损失函数，反向传播
        loss = loss_function(output, label)
        loss.backward()

        optimizer.step()                  # 优化器迭代一步更新参数
```

这里注意的就是要在回传梯度前调用 optimizer.zero_grad()清空参数原有梯度，并在反向传播求得梯度后，调用 optimizer.step()更新参数。

当有时训练过程较长，或在训练中途程序进程被停止时，如果再重新从头训练会花费很多时间。但是如果能够将优化器中途状态保留并存储下来，并再继续以此为基础进行训练，则可以没有损失。optimizer 里提供了.state_dict()和.load_state_dict()两种方法，分别用来导出优化器此时的中间状态，并加载已有状态。这种方法在 nn.Module 里也有，是为了保存和加载预训练模型。我们将在第 4 章中进一步介绍模型存储和加载方法。

### 3.2.2 经典优化器介绍

这一节我们介绍一些常见的优化算法及其优化器。首先最经典常用的是 optim.SGD 优化器，实现了随机梯度下降法。参数列表为(params, lr, momentum, dampening, weight_decay, nesterov)，分别是模型待训练参数、学习率、动量因子、动量阻尼系数、权重衰减、是否启用 Nesterov 动量加速。一般来说调整学习率、动量因子和权重衰减这些超参数。学习率控制参数更新步长，动量因子控制每次更新相较于之前的比例，权重衰减相当于增加权重的正则项提高模型泛化能力。这些参数往往需要适当调整，最好是基于已有文献中的结果进行调整。其更新权重的经典算法[30]为：

$$v_{t+1} = \mu * v_t + lr * g_{t+1}$$

$$p_{t+1} = p_t - v_{t+1}$$

其中 $p$, $g$, $v$, $\mu$ 分别为参数、梯度、速度和动量。注意 PyTorch 中的实现与此略有不同，具体差异参见官方文档说明[①]。

另一个重要的优化器为 optim.Adam，实现了论文[31]中的优化算法。但是在理解 Adam 算法之前，先理解另一个优化算法 RMSprop（Root Mean Square Propagation）比较有帮助。RMSprop 是在 Geoffrey Hinton 的线上神经网络课程[②]中提出的。其主要思想是原 SGD 算法中对于每个参数的更新都使用了同样的学习率，而实际上可以根据权重的大小进行调整。其更新算法为：

$$v_{t+1} = \gamma * v_t + (1 - \gamma) * (\nabla_w L)^2$$
$$p_{t+1} = p_t - \frac{\eta}{\sqrt{v_{t+1}} + \epsilon} \nabla_w L$$

其中 $\nabla_w L$ 为梯度。可以看出主要变化在于滑动平均统计了梯度的平方项，并以此调整更新步长。而 Adam 算法则更进一步改进以上更新算法，同时滑动平均统计梯度及其平方项。

$$m_t = \beta_1 m_{t-1} + (1 - \beta_1) \nabla_w L$$
$$v_t = \beta_2 v_{t-1} + (1 - \beta_2) (\nabla_w L)^2$$
$$\widehat{m_t} = \frac{m_t}{1 - \beta_1^t}, \quad \widehat{v_t} = \frac{v_t}{1 - \beta_2^t}$$
$$p_t = p_{t-1} - \frac{\eta}{\sqrt{\widehat{v_t}} + \epsilon} \widehat{m_t}$$

optim.Adam 的参数列表为(params, lr, betas, eps, weight_decay)，其中 betas 和 eps 就是上式中的 $\beta_1, \beta_2, \epsilon$。而在最近有针对 Adam 算法中的权重衰减部分的细致研究发现原始实现中的错误，并改进提出了新的 AdamW 算法[32]，也在 optim.AdamW 中实现了。更多的优化器类型总结在表 3.3 中。

表 3.3 优化器类型

| Adadelta | AdamW | ASGD | Rprop |
| Adagrad | SparseAdam | LBFGS | SGD |
| Adam | Adamax | RMSprop | |

### 3.2.3 学习率调整

在训练模型时，还有另外一项重要的步骤对最后模型性能起到重要作用，便是学习率的调整。之前介绍的多种优化器默认是按照一个初始学习率，一直进行权重更新。而实际上在实验中发现，在训练过程中降低学习率大小，有助于模型在进入性能平台期之

---

① https://pytorch.org/docs/stable/optim.html#torch.optim.SGD

② https://www.cs.toronto.edu/~tijmen/csc321/slides/lecture_slides_lec6.pdf

后，进一步提高准确性能。因此在 PyTorch 1.0 版本之后，增加了 optim.lr_scheduler 模型，提供多种调整学习率的方法。一个学习率调整器的常见使用方法为：

```
scheduler = optim.lr_scheduler.SomeScheduler(optimizer, *args)
for epoch in range(NUM_EPOCHS):
    train(...)
    test(...)
    scheduler.step()
```

即按照一次训练一次测试周期后，调整优化器中的学习率。

下面介绍几种常见的学习率调整方法。首先是 optim.lr_scheduler.StepLR，表示按照一定的周期间隔衰减学习率。其初始化参数列表为(optimizer, step_size, gamma, last_epoch)，其中 optimizer 为待调整的优化器，step_size 表示间隔周期长，gamma 为每次衰减比例。而有些时候并非固定间隔衰减，因此提供了 optim.lr_scheduler.MultiStepLR，指定在哪些周期时衰减，更加的灵活。其初始化参数列表为(optimizer, milestones, gamma, last_epoch)，其中 milestones 就是指定各个周期数。比如总训练周期为 160，在第 80 周期和第 120 周期进行 0.1 比例衰减，则可以写为 optim.lr_scheduler.MultiStepLR(optimizer, milestones=[80, 120], gamma=0.1)。还有最近研究提出的按照余弦衰减[33]（cosine annealing）的方式调整学习率。其调整规则为：

$$\eta_t = \eta_{min} + \frac{1}{2}(\eta_{max} - \eta_{min})\left(1 + \cos\left(\frac{T_{cur}}{T_{max}}\pi\right)\right)$$

图 3.3 中以初始学习率为 0.1，总训练周期数为 100，展示了常见的学习率调整算法产生的可视化结果。已有的学习率调整器类型总结在表 3.4 中。

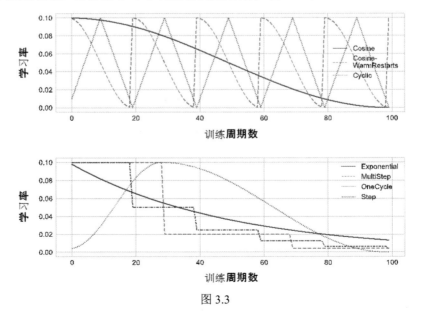

图 3.3

表 3.4　学习率调整器类型

| LambdaLR | MultiStepLR | ReduceLROnPlateau | OneCycleLR |
| --- | --- | --- | --- |
| MultiplicativeLR | ExponentialLR | CosineAnnealingLR | |
| StepLR | CyclicLR | CosineAnnealingWarmRestarts | |

## 3.3　PyTorch 应用实战一：实现二值化神经网络

在这一节中我们将综合利用之前所介绍的 PyTorch 构建神经网络的核心模块 torch.nn 以及优化模块 torch.optim，来实现一个二值化神经网络的训练。相比于标准的基于浮点数运算的神经网络，二值化神经网络将内部主要运算化简为+1/-1 的二值操作，可以起到节省内存加速的目的。接下来我们就介绍经典的二值化网络 BinaryNet，以及如何实现并训练。

### 3.3.1　二值化网络 BinaryNet 概述

BinaryNet 首次提出于论文 "Binarized Neural Network"[34]之中，其主要想法是将原神经网络中的权重响应均变为只由+1/-1 表示的二值情形，则此时的乘法运算可以简化成为 popcount 位运算，减少了内存占用，加速了运算过程。其二值化操作也是很直观，对于一个浮点值 Tensor，无论其为权重还是响应，其二值化结果为：

$$x^b = \text{sign}(x) = \begin{cases} +1, & x \geq 0 \\ -1, & x < 0 \end{cases}$$

虽然二值化方式很简单，但是这引起了训练上的困难，因为 sign 函数是离散输出不可导的，无法直接实现自动反向传递导数。这里我们将采用在第 2.2.4 节中介绍过的方法，利用 PyTorch 自动求导的 Function 类实现特定的反向传播过程。在原论文中使用了如下的梯度回传过程：

$$\frac{\partial \text{loss}}{\partial x} = \frac{\partial \text{loss}}{\partial x^b} * \mathbb{I}(|x| \leq 1)$$

即损失函数对于浮点值 $x$ 的梯度等于对于二值化输出 $x^b$ 的梯度，只不过仅在 $|x| \leq 1$ 的区间成立，当 $x$ 的绝对值大于 1 时，梯度为 0。这种做法对于 $x$ 较小时可以正常训练，当 $x$ 较大时截止其回传梯度，防止进一步数值增大。

以上的做法解决了主要运算层（如线性层、卷积层）的二值化过程。但是此时的运算层输出相当于两个二值张量乘法和累加结果，往往会导致输出异常变大，为此需要引入 BatchNorm 层将其归一化。而在归一化后的输出变成了浮点数，因此在进入下一个线性层运算之前，将输入进行二值化即可。

最后在更新参数时，原文中提到一个技巧。由于我们在反向传播梯度时，截止了输入绝对值大于 1 的部分，而这一部分参数是没有接收到梯度更新的，因此也不会继续改变。因此在梯度回传更新参数后，可以强制截止参数至[-1, 1]区间，避免参数进一步累积扩大。

### 3.3.2 具体实现

这里我们采用类似于第 2.4 节中的做法，在 MNIST 数据集上进行数字分类实验。首先是实现二值化的线性层：

```python
import torch
import torch.nn as nn
import math
from torch.autograd import Function
import torch.nn.functional as F
from torchvision import datasets, transforms


class Binarize(Function):
    @staticmethod
    def forward(ctx, input):                       # 前向运算过程
        ctx.save_for_backward(input)
        return torch.sign(input + 1e-20)           # 避免出现输入为 0 无法判断符号

    @staticmethod
    def backward(ctx, grad_output):                # 反向运算过程
        input = ctx.saved_tensors[0]
        grad_output[input > 1] = 0
        grad_output[input < -1] = 0
        return grad_output


class BinarizedLinear(nn.Module):
    def __init__(self,
                 in_features,
                 out_features,
                 binarize_input=True):             # 控制输入是否二值化
        super(BinarizedLinear, self).__init__()
        self.binarize_input = binarize_input
        self.weight = nn.Parameter(
            torch.Tensor(out_features, in_features)
        )
        nn.init.kaiming_uniform_(                  # 权重初始化
            self.weight, a=math.sqrt(5)
        )

    def forward(self, x):
        if self.binarize_input:                    # 二值化输入
            x = Binarize.apply(x)
        w = Binarize.apply(self.weight)            # 二值化权重
        out = torch.matmul(x, w.t())
```

```
        return out
```

接下来是模型搭建和优化器设置部分。这里模型时一个包含两个因此的 MLP，中间加入了 Dropout 层增强泛化能力。优化器使用了 Adam 优化算法，学习率调整器使用了 StepLR，按照每 40 个周期下降为原来的十分之一。

```
model = nn.Sequential(
    BinarizedLinear(784, 2048, False),      # 第一层输入不进行二值化
    nn.BatchNorm1d(2048),
    BinarizedLinear(2048, 2048),
    nn.BatchNorm1d(2048),
    BinarizedLinear(2048, 2048),
    nn.Dropout(0.5),                         # 加入 Dropout 增强泛化性
    nn.BatchNorm1d(2048),
    nn.Linear(2048, 10)
)

optimizer = torch.optim.Adam(                # Adam 优化器
    model.parameters(), lr=LR
)
scheduler = torch.optim.lr_scheduler.StepLR( # StepLR 学习率调整器
    optimizer, step_size=40, gamma=0.1
)
```

然后是常规的数据加载，训练测试部分。

```
TRAIN_BATCH_SIZE = 64
TEST_BATCH_SIZE = 1000
LR = 0.01
EPOCH = 100
LOG_INTERVAL = 100

train_loader = torch.utils.data.DataLoader(  # 训练数据
    datasets.MNIST(
        './data', train=True, download=True,
        transform=transforms.Compose([
            transforms.ToTensor(),
            transforms.Normalize((0.1307,), (0.3081,))
        ])
    ),
    batch_size=TRAIN_BATCH_SIZE, shuffle=True)

test_loader = torch.utils.data.DataLoader(   # 测试数据
    datasets.MNIST(
        './data', train=False,
        transform=transforms.Compose([
            transforms.ToTensor(),
            transforms.Normalize((0.1307,), (0.3081,))
        ])
    ),
```

```python
            batch_size=TEST_BATCH_SIZE, shuffle=False)

for epoch in range(EPOCH):
    for idx, (data, label) in enumerate(train_loader):
        model.train()                                    # 设置模型为训练模式
        optimizer.zero_grad()
        output = model(data.view(-1, 28*28))             # 输入转换为784维向量
        loss = F.cross_entropy(output, label)
        loss.backward()

        optimizer.step()                                 # 更新参数并截止到[-1, 1]区间
        for p in model.parameters():
            p.data.clamp_(-1, 1)

        if idx % LOG_INTERVAL == 0:
            print('Epoch %03d [%03d/%03d]\tLoss: %.4f' % (
                epoch, idx, len(train_loader), loss.item()
            ))

    correct_num = 0
    total_num = 0
    with torch.no_grad():                                # 测试阶段
        for data, label in test_loader:
            model.eval()                                 # 设置模型为测试模式
            output = model(data.view(-1, 28*28))
            pred = output.max(1)[1]
            correct_num += (pred == label).sum().item()
            total_num += len(data)

    acc = correct_num / total_num                        # 计算准确率
    print('...Testing @ Epoch %03d\tAcc: %.4f' % (
        epoch, acc
    ))

    scheduler.step()                                     # 利用学习率调整器更新学习率
```

最后运行程序，如果一切正常将会输出训练信息和测试信息如下：

```
...Testing @ Epoch 097  Acc: 0.9832
Epoch 098 [000/938]     Loss: 0.0000
Epoch 098 [100/938]     Loss: 0.0207
Epoch 098 [200/938]     Loss: 0.0000
Epoch 098 [300/938]     Loss: 0.0000
Epoch 098 [400/938]     Loss: 0.0001
Epoch 098 [500/938]     Loss: 0.0001
Epoch 098 [600/938]     Loss: 0.0000
Epoch 098 [700/938]     Loss: 0.0004
Epoch 098 [800/938]     Loss: 0.0000
```

```
Epoch 098 [900/938]      Loss: 0.0000
...Testing @ Epoch 098   Acc: 0.9829
Epoch 099 [000/938]      Loss: 0.0000
Epoch 099 [100/938]      Loss: 0.0001
Epoch 099 [200/938]      Loss: 0.0084
Epoch 099 [300/938]      Loss: 0.0000
Epoch 099 [400/938]      Loss: 0.0012
Epoch 099 [500/938]      Loss: 0.0002
Epoch 099 [600/938]      Loss: 0.0006
Epoch 099 [700/938]      Loss: 0.0000
Epoch 099 [800/938]      Loss: 0.0000
Epoch 099 [900/938]      Loss: 0.0006
...Testing @ Epoch 099   Acc: 0.9826
```

可以看出最终的测试准确率水平在 98.3%左右，这和我们在第 2.4 节中所实现的网络预测水平接近，也说明了二值化网络即使以二值运算代替浮点数运算也可以保证预测精度不损失，拥有巨大的应用前景。

## 3.4 PyTorch 应用实战二：利用 LSTM 实现文本情感分类

在这一节中我们将继续利用 PyTorch 中的 nn.Module 模块构建一个文本情感分类任务。由于文本属于序列数据，因此将采用反馈神经网络中的 LSTM 模型实现输入特征提取。这一节中将会展示如何在 PyTorch 中处理文本类数据做法。

### 3.4.1 文本情感分类

所谓文本情感分析（text sentiment analysis），指的是给定一段文本对其中所包含的语义态度进行分析。而文本情感分类则将这些语义情感分为离散的多种类别，将此问题转化为分类问题。此次我们将要进行实验的数据集是 Stanford Sentiment Treebank（SST）数据集，其中包含了从众多电影评论中提取的 20 多万条语句的情感态度标签，一个有 5 种，分为非常负面、负面、中性、正面和非常正面。下面展示了一些数据集中的样本和对应的标签：

```
[very negative]: I simply cannot recommend it enough.

[negative]: Starts out ballsy and stylish but fails to keep it up and settles
into clichés.

[neutral]: Anyone who gets chills from movies with giant plot holes will
find plenty to shake and shiver about in `The Ring.'

[positive]: One of the most exciting action films to come out of China in
recent years.

[very positive]: That Jack Nicholson makes this man so watchable is a tribute
```

```
not only to his craft , but to his legend.
```

一般对于文本信息需要进行预处理，使得其从文本符号转换为可被神经网络处理的数值张量。这些预处理操作可大致分为将文本分割成为 token（tokenization），建立词表（vocabulary），将文本根据词表转换为索引向量，再利用词嵌入表查找（embedding lookup）转换为最终的数值向量表示。以下例子展示了上述过程：

```
"I simply can't recommend it enough."
--> tokenization
["I", "simply", "cannot", "recommend", "it", "enough", "."]

--> vocabulary
{"I": 0, "simply": 1, "cannot": 2, ..., ".": 6}

--> indexing
[0, 1, 2, 3, 4, 5, 6]

--> embedding lookup
[[ 8.7705e-01, -1.0287e+00, ...,  3.5242e-01,  3.6361e-01,  1.6073e-01],
 [ 1.3670e-01,  2.2594e-01, ..., -2.2748e-01,  3.2125e-02, -1.3373e-01],
 ...,
 [-2.5318e-01,  1.2764e-01, ..., -7.8879e-01,  1.0301e-01, -1.8828e-01],
 [-1.2559e-01,  1.3630e-02, ..., -3.4224e-01, -2.2394e-02,  1.3684e-01]]
```

这样一个长度为 $L$ 的文本字符串，就转化成了形状为 $L\times N_e$ 的张量，其中 $N_e$ 为词嵌入的维度。而对于 batch size 为 $B$ 的一组文本字符串，在 PyTorch 中最后转化成的输入张量形状为 $L\times B\times N_e$，即第二维才是 batch size，这一点需要注意。对于长度不同的文本，补零到同一长度。而这个词嵌入表则是通过著名的词向量预训练算法得到的。此处我们使用 "GloVe: Global Vectors for Word Representation"[35]论文中产生的 glove.6B.300d 词向量表，表示其包含 60 亿 token 的 300 维词向量。

以上的各个步骤环节需要复杂的处理流程。但是好在 PyTorch 官方提供了 torchtext[①]这个库，集合了常见的自然语言处理中使用的数据集，并且提供了简洁的接口用以产生预处理后的 dataloader。这一部分内容我们将在下一节中代码进行介绍。

另一方面，在解决了数据输入方面的问题，我们需要考虑模型设计。这里我们使用了多层 LSTM 的反馈神经网络。在第 3.1.3 节中我们也介绍过，nn.LSTM 实现了多层 LSTMCell 的堆叠，使得每一个反馈 Cell 的输出是下一个的输入，构成深层模型。而在此处文本情感分类任务中，需要将一个序列输入转换为一个标签类别预测结果，因此通常直接的做法是取最高层反馈 Cell 的最后一个时刻的特征输出。然后再通过线性层转换成类别预测。更加复杂的做法可以引入注意力（Attention）机制，将多个时刻的特征综合在一起。这部分留待读者自己探索。

---

① https://torchtext.readthedocs.io/en/latest/

## 3.4.2 具体实现

接下来介绍具体的实现过程。首先是参数设置和数据准备阶段。这里利用 torchtext 完成上述的文本预处理。在安装 torchtext 时，还需要安装相关依赖的 nltk[①]库。均可以通过 pip install 安装完成。

```
import torch
import torch.optim as optim
import torch.nn as nn
import numpy as np
import torch.nn.functional as F
from torchtext import data, datasets

class Args:                                      # 参数设置
    max_vocab_size = 25000                       # 限定词表最大规模
    n_labels = 5                                 # 情感类别个数
    epochs = 5                                   # 训练总周期
    embedding_dim = 300                          # 词向量维度
    hidden_dim = 512                             # LSTMCell 隐层维度
    n_layers = 3                                 # LSTM 层数
    batch_size = 64                              # 批样本量大小
    display_freq = 50                            # 输出信息间隔
    lr = 0.01                                    # 学习率

args = Args()

TEXT = data.Field()                              # TEXT, LABEL 用以处理文本和标签
LABEL = data.LabelField(dtype=torch.float)

train_data, valid_data, test_data = datasets.SST.splits(
    TEXT, LABEL, fine_grained=True               # 分割 SST 数据集
)                                                # fine_grained 指定情绪分类为 5 类

TEXT.build_vocab(                                # 构建词表
    train_data,
    max_size=args.max_vocab_size,
    vectors="glove.6B.300d",                     # 对应的词向量表
    unk_init=torch.Tensor.normal_                # 不在词向量表中的初始化方式
)

LABEL.build_vocab(train_data)                    # 构建标签

device = 'cpu'                                   # 设置运行在 CPU 上

train_iter, valid_iter, test_iter = data.BucketIterator.splits(
```

---

[①] https://www.nltk.org

```
    (train_data, valid_data, test_data),    # 产生 dataloader
    batch_size=args.batch_size,
    device=device
)
```

接下来是构建模型和优化器部分：

```
input_dim = len(TEXT.vocab)
output_dim = args.n_labels

class Model(nn.Module):
    def __init__(self,
                 in_dim,
                 emb_dim,
                 hid_dim,
                 out_dim,
                 n_layer):

        super(Model, self).__init__()
        self.embedding = nn.Embedding(in_dim, emb_dim)
        self.rnn = nn.LSTM(emb_dim, hid_dim, n_layer)
        self.linear = nn.Linear(hid_dim, out_dim)
        self.n_layer = n_layer
        self.hid_dim = hid_dim

    def forward(self, text):
        # text = [sentence len, batch size]

        embedded = self.embedding(text)        # 获取向量表示
        # embedded = [sentence len, batch size, emb dim]

        h0 = embedded.new_zeros(               # 隐状态初始化
            self.n_layer, embedded.size(1), self.hid_dim
        )
        c0 = embedded.new_zeros(
            self.n_layer, embedded.size(1), self.hid_dim
        )
        output, (hn, cn) = self.rnn(embedded, (h0, c0)) # LSTM 产生输出
        # output: [sentence len, batch size, hid dim]

        return self.linear(output[-1])         # 转换为类别预测输出

model = Model(input_dim,
              args.embedding_dim,
              args.hidden_dim,
              output_dim,
              args.n_layers)

pretrained_embeddings = TEXT.vocab.vectors  # 将 embedding 用词向量初始化
model.embedding.weight.data.copy_(pretrained_embeddings)
```

```python
model.to(device)                                    # 模型迁移

optimizer = optim.Adam(                             # 构建优化器
    model.parameters(), lr=args.lr
)
```

最后便是训练和测试部分。这里与之前 MNIST 数字分类稍有差别的地方在于，由于存在着验证集和测试集之间的区分，所以需要在验证集上选取多次分类准确率最高的模型，再在测试集上测试。

```python
def train(epoch, model, iterator, optimizer):       # 训练函数
    loss_list = []
    acc_list = []
    
    model.train()                                   # 设置为训练状态
    
    for i, batch in enumerate(iterator):
        optimizer.zero_grad()
        text = batch.text.to(device)
        label = batch.label.long().to(device)
        predictions = model(text)
        loss = F.cross_entropy(predictions, label)
        loss.backward()
        optimizer.step()
        
        acc = (predictions.max(1)[1] == label).float().mean()
        loss_list.append(loss.item())               # 累积 loss 和 acc 值
        acc_list.append(acc.item())
        
        if i % args.display_freq == 0:
            print("Epoch %02d, Iter [%03d/%03d], "
                "train loss = %.4f, train acc = %.4f" % (
                epoch, i, len(iterator),
                np.mean(loss_list), np.mean(acc_list)
            ))
            loss_list.clear()                       # 报告平均值后清空,再重新记录
            acc_list.clear()


def evaluate(epoch, model, iterator):               # 测试函数
    val_loss = 0
    val_acc = 0
    
    model.eval()                                    # 设置为验证状态
    
    with torch.no_grad():                           # 关闭自动求导
        for batch in iterator:
            text = batch.text.to(device)
```

```python
            label = batch.label.long().to(device)
            predictions = model(text)
            loss = F.cross_entropy(predictions, label)
            
            acc = (predictions.max(1)[1] == label).float().mean()
            val_loss += loss.item()
            val_acc += acc.item()
        
        val_loss = val_loss / len(iterator)
        val_acc = val_acc / len(iterator)
        print('...Epoch %02d, val loss = %.4f, val acc = %.4f' % (
            epoch, val_loss, val_acc
        ))
        return val_loss, val_acc                  # 报告损失函数和准确率

best_acc = 0
best_epoch = -1
for epoch in range(1, args.epochs + 1):
    train(epoch, model, train_iter, optimizer)
    valid_loss, valid_acc = evaluate(epoch, model, valid_iter)
    if valid_acc > best_acc:                     # 保留验证集上准确率最高的模型
        best_acc = valid_acc
        best_epoch = epoch
        torch.save(                              # 存储模型并命名为 best-model.pth
            model.state_dict(),
            'best-model.pth'
        )

print('Test best model @ Epoch %02d' % best_epoch)
model.load_state_dict(torch.load('best-model.pth'))    # 加载最好的模型
test_loss, test_acc = evaluate(epoch, model, test_iter)
print('Finally, test loss = %.4f, test acc = %.4f' % (
    test_loss, test_acc                          # 报告最终测试集上的准确率
))
```

其中程序运行之初需要下载 SST 数据集和大约 800MB 大小的 GloVe 词向量，因此会花费一些时间。如果可以拥有 GPU 计算资源，也可以将以上模型数据迁移到 GPU 上计算实现加速，只需要将原代码中的 device='cuda' 即可。如果一切运行正常，将输出以下信息。最终可以实现在测试集上大约 30%左右的准确率。

```
$ python sentiment_analysis.py
Epoch 01, Iter [000/134], train loss = 1.6094, train acc = 0.2031
Epoch 01, Iter [050/134], train loss = 1.6470, train acc = 0.2637
Epoch 01, Iter [100/134], train loss = 1.5944, train acc = 0.2603
...Epoch 01, val loss = 1.5904, val acc = 0.2549
Epoch 02, Iter [000/134], train loss = 1.5643, train acc = 0.3594
Epoch 02, Iter [050/134], train loss = 1.5820, train acc = 0.2709
Epoch 02, Iter [100/134], train loss = 1.5922, train acc = 0.2603
```

```
...Epoch 02, val loss = 1.5729, val acc = 0.2913
Epoch 03, Iter [000/134], train loss = 1.6261, train acc = 0.1562
Epoch 03, Iter [050/134], train loss = 1.5706, train acc = 0.2812
Epoch 03, Iter [100/134], train loss = 1.5766, train acc = 0.2641
...Epoch 03, val loss = 1.5749, val acc = 0.2524
Epoch 04, Iter [000/134], train loss = 1.6339, train acc = 0.2188
Epoch 04, Iter [050/134], train loss = 1.5726, train acc = 0.2812
Epoch 04, Iter [100/134], train loss = 1.5770, train acc = 0.2744
...Epoch 04, val loss = 1.5944, val acc = 0.3166
Epoch 05, Iter [000/134], train loss = 1.4602, train acc = 0.3281
Epoch 05, Iter [050/134], train loss = 1.5750, train acc = 0.2884
Epoch 05, Iter [100/134], train loss = 1.5711, train acc = 0.2762
...Epoch 05, val loss = 1.6176, val acc = 0.3036
Test best model @ Epoch 04
...Epoch 05, val loss = 1.6409, val acc = 0.3145
Finally, test loss = 1.6409, test acc = 0.3145
```

# 第 4 章　基于 PyTorch 构建复杂应用

在前面的内容中，我们已经探索了 PyTorch 中的核心部分，包括张量计算和神经网络模块。理论上来说依赖这些核心功能就可以完成一般的应用任务。然而当实际进行复杂应用构建时，其中涉及的数据预处理、数据加载流程、复杂模型搭建、模型存储及转换以及实验过程中的日志记录等等方面，还是需要深入探索。PyTorch 也非常贴心地实现了以上功能，避免了烦琐的操作流程。本章我们将介绍 PyTorch 的这些功能，并在最后应用实战环节进一步实现具有挑战性的真实任务。

## 4.1　PyTorch 数据加载

在前面的实战环节中，关于数据预处理和数据加载部分没有作过多介绍，仅是将其作为标准模板代码使用。然而数据加载部分往往是真实应用任务中至关重要的环节。其中的数据预处理可以进行数据增广（Data Augmentation），对训练模型预测性能有着显著的提升。而其中的数据加载可以对大规模数据集进行异步式处理，从而避免模型整体训练时间因数据吞吐缓慢而产生延时。在这一节中，将深入其中的运行机制，从而实现更多复杂的数据加载流程。

### 4.1.1　数据预处理：torchvision.transforms

首先我们介绍 PyTorch 针对计算机视觉任务中的数据预处理模块 torchvision.transforms。一个常见的数据预处理流程可以写作如下形式：

```
transform_train = transforms.Compose([
    transforms.RandomCrop(32, padding=4),
    transforms.RandomHorizontalFlip(),
    transforms.ToTensor(),
    transforms.Normalize((0.4914, 0.4822, 0.4465),
                         (0.2023, 0.1994, 0.2010)),
])
```

其中 transforms.Compose 是组合操作，可以将多个图像变换操作组合在一起，构成一个预处理流程。其接受参数为列表，包含多个图像变换操作。transforms.RandomCrop 和 transforms.RandomHorizontalFilp 都是图像处理操作，一个是随机在图片上裁切（crop）出来一定大小的图片，参数列表是 RandomCrop(image_size, padding)，分别制定裁切大小和原图周围补零个数；另一个随机对图片进行水平翻转操作。这一类图像变换操作都是对于 PIL 图片对象。PIL（Python Image Library）是 PyTorch 的图像处理标准库。使用前可以通过 pip install pillow 安装。以上两个图像变换操作产生的结果可以通过如下代码展示：

```
>>> import torchvision.transforms as tfm
>>> from PIL import Image
>>> img = Image.open('cat.jpg')                          # 读取图片
>>> img_1 = tfm.RandomCrop(200, padding=50)(img)         # 随机裁切图片
>>> img_1.show()                                          # 展示图片
>>> img_1.save('crop.jpg')                                # 保存图片
>>> img_2 = tfm.RandomHorizontalFlip()(img)              # 随机水平翻转图片
>>> img_2.show()
>>> img_2.save('flip.jpg')
```

展示的结果如图 4.1 所示。

原图　　　　　　　　随机裁切图　　　　　　　水平翻转图

图 4.1

那么以上的图片变换内部实现机制又是怎样的呢？我们可以以 HorizontalFlip 为例，观察其实现的源码。可以看出每一个图像变换在初始化参数后，即为一个可调用对象，实现 __call__ 中的操作。

```
class RandomHorizontalFlip(object):
    """Horizontally flip the given
       PIL Image randomly with a given probability.
    Args:
        p (float): probability of the image
            being flipped. Default value is 0.5
    """
    def __init__(self, p=0.5):
        self.p = p

    def __call__(self, img):
        """
        Args:
            img (PIL Image): Image to be flipped.
        Returns:
            PIL Image: Randomly flipped image.
        """
        if random.random() < self.p:
            return F.hflip(img)
        return img
```

```
    def __repr__(self):
        return self.__class__.__name__ + '(p={})'.format(self.p)
```

而上面 F.hflip 函数的具体实现又是如下代码所示，本质上还是调用了 PIL 中的图片操作运算。因此我们实际上可以自由构建图像变换操作，然后在数据预处理流程中调用。

```
def hflip(img):
    """Horizontally flip the given PIL Image.
    Args:
        img (PIL Image): Image to be flipped.
    Returns:
        PIL Image:  Horizontall flipped image.
    """
    if not _is_pil_image(img):
        raise TypeError(
          'img should be PIL Image. Got {}'.format(type(img))
        )
    return img.transpose(Image.FLIP_LEFT_RIGHT)
```

更多的图像变换操作展示在图 4.2 中，每张图片下方显示对应的变换方法。

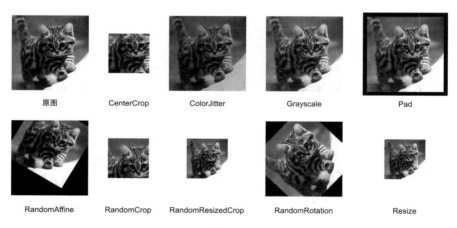

图 4.2

在一开始的数据处理流程中，还包含另外两个操作 transforms.ToTensor 和 transforms.Normalize,其中 ToTensor 是转换操作,将输入的图片转换为 PyTorch 中的 Tensor。这个操作还可以对 numpy.ndarray 格式的输入转换为 Tensor。注意此处转换是将一个形状为 $H \times W \times C$ 的输入转换为 $C \times H \times W$ 形状的张量。同时还将输入数值范围$[0, 255]$归一化到$[0, 1]$。如果想进行逆变换，将 Tensor 转换为图片，则使用 transforms.ToPILImage 操作。transforms.Normalize 则是在 Tensor 上进一步变换，利用给定的均值和标准差对输入进行标准化，初始化方式为 transforms.Normalize(mean, std),其中 mean 和 std 都是逐通道的均值和方差。完整的变换操作类型总结在表 4.1 中。

表 4.1 变换操作类型

| Compose | RandomAffine | RandomOrder | Resize | ToTensor |
| --- | --- | --- | --- | --- |
| CenterCrop | RandomApply | RandomPerspective | Scale | Lambda |
| ColorJitter | RandomChoice | RandomResizedCrop | TenCrop | |
| FiveCrop | RandomCrop | RandomRotation | LinearTransformation | |
| Grayscale | RandomGrayscale | RandomSizedCrop | Normalize | |
| Pad | RandomHorizontalFlip | RandomVerticalFlip | ToPILImage | |

其实数据预处理操作不仅限于以上几种。这里推荐 albumentations[①]这个项目，其中包含了大量额外的图像变换操作，同时还包括了针对物体检测、图像分割等任务中特有的数据预处理流程。图 4.3 中展示了其中可实现的多种变换操作。

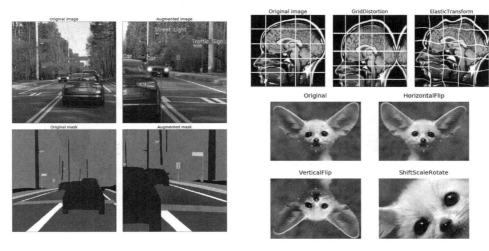

图 4.3

同时最近的多项研究探索了数据预处理流程的自动化实现问题。我们在一开始所展示的预处理流程是按照经验人为设置进行的组合。而在 Google Brain 团队则提出了 AutoAugment[34]算法，通过遗传进化算法迭代式地筛选出最优的数据预处理流程，并且显著提升了训练模型的预测性能。可以看出即使是看似平凡普通的数据预处理阶段，也是值得深入研究探索，大有可为。

### 4.1.2 数据加载：torch.utils.data

前面介绍了数据预处理，接下来就是如何加载数据构建输入流。首先是要实现如何加载数据，这里使用 torch.utils.data.Dataset 类。PyTorch 支持两种模式的数据读取，一种

---

① https://github.com/albumentations-team/albumentations

是映射式（map-style），一种是迭代式（iterable-style）。对于映射式数据读取，其实现形式如下：

```python
import torch.utils.data as data

class MyDataset(data.Dataset):
    def __init__(self, data):
        self.data = data                    # 数据集参数初始化阶段
        ...

    def __getitem__(self, index):           # 根据索引获取单个数据点
        ...
        return self.data[index]

    def __len__(self):                      # 获取数据集总样本数量
        return len(self.data)
```

这种形式的数据集实现了单个索引指标和对应数据点之间的映射。因此只需要关注单个数据的变换操作，之后会有 torch.uitls.data.DataLoader 来负责多线程读取和数据点合并。这种适合数据集可全部读入内存或者硬盘支持频繁快速的随机读取。比如像配备了固态硬盘的设备，即使是在面对如 ImageNet 这种百万量级图片的数据规模时，也可以用映射式数据集，因为只需要提前加载每个样本的路径，而在__getitem__方法内部实现随机读取。另一种迭代是数据集更多是面对如实时数据流进来的顺序数据，或者设备不支持频繁随机读取，只能以连续块状读取数据。此时则需要使用 data.IterableDataset 类，并实现其__iter__方法，决定每一次迭代返回的数据。对于一些常见图像分类，物体检测，语义分割，视频分类等数据集，torchvision.dataset 模块已经实现可以直接调用，所实现的数据集名称总结在表 4.2 中。

表 4.2　数据集名称

| MNIST | CocoCaptions | CIFAR10 | Flickr8k | USPS |
|---|---|---|---|---|
| FashionMNIST | CocoDetection | CIFAR100 | Flickr30k | Kinetics400 |
| KMNIST | LSUN | STL10 | VOCSegmentation | HMDB51 |
| EMNIST | ImageFolder | SVHN | VOCDetection | UCF101 |
| QMNIST | DatasetFolder | PhotoTour | Cityscape | CelebA |
| FakeData | ImageNet | SBU | SBDataset | |

实现了 Dataset 类只是解决了数据读取的这一步，而 torch.utils.data.DataLoader 则是完整实现了提供数据流水线（data pipeline）。一个常见的 DataLoader 写法如下：

```
loader = torch.utils.data.DataLoader(
    dataset, batch_size=32, shuffle=True, sampler=None,
    num_workers=2, collate_fn=None, pin_memory=True, drop_last=False
)
```

其中涉及到多个参数。首先 dataset 是传入 DataLoader 作为其构造函数的基础，即上面所介绍的几种 Dataset。batch_size 则是指定批样本量大小。shuffle 为布尔变量，指定该 DataLoader 是否在每次迭代时打乱数据点的顺序，常用在训练阶段产生随机数据集。接下来几个参数设置需要单独详细介绍。

首先是 sampler 的设置。sampler 的目的是在每次迭代时产生一组索引。这组索引供给于 Dataset 获取每个数据点，最后再后续产生 data pipeline。而 sampler 的设置和 shuffle 参数设置是互斥的。就是当 shuffle 设置为 True 时，不可以再设置 sampler，因为此时已经默认调用其内部实现的 RandomSampler。以典型的 SequentialSampler 类的实现展示其工作方式如下：

```python
class SequentialSampler(Sampler):
    """Samples elements sequentially, always in the same order.

    Arguments:
        data_source (Dataset): dataset to sample from
    """

    def __init__(self, data_source):
        self.data_source = data_source

    def __iter__(self):
        return iter(range(len(self.data_source)))

    def __len__(self):
        return len(self.data_source)
```

可以看出核心就是实现其 __iter__ 方式，决定每次迭代返回的索引指标。PyTorch 中内置了 5 种 sampler，它们是 SequentialSampler、RandomSampler、SubsetRandomSampler、WeightedRandomSampler 和 BatchSampler，分别对应顺序采样、随机采样、子集随机采样、加权随机采样和批采样。顺序采样用于 shuffle=False 时，随机采样用于 shuffle=True 时。子集随机采样用于将原始整个数据划分成为训练集和验证集。使用方式如下：

```python
>>> from torch.utils.data import SubsetRandomSampler
>>> indices = list(range(20))
>>>
>>> sampler = SubsetRandomSampler(indices[:10])    # 只取前 10 个索引
>>> print([i for i in sampler])
[9, 1, 6, 0, 8, 5, 4, 2, 3, 7]
```

如果想要在运行时在线变更 indices，则可以原地修改 sampler 中的 indices 成员变量来实现：

```python
>>> sampler.indices.clear()                              # 清空原来索引列表
>>> sampler.indices.extend(list(range(20, 30)))          # 变更新索引列表
>>> print([i for i in sampler])
```

```
[22, 25, 28, 24, 29, 26, 21, 27, 20, 23]
```

加权随机采样则对每个索引按照不同概率进行采样，比如对于每次训练分类错误的样本进行更多权重采样，提升分类效果。最后批采样（BatchSampler）则是封装以上单点数据的采样器，根据 DataLoader 中设定的 batch_size 将多个样本索引组合成 batch 返回。其中有一个 drop_last，当设置为 True 时，则当最后返回的 batch 索引数量小于 batch size 时将其抛弃。

其次是 num_worker 的设置，其服务于多进程（multiprocessing）数据加载，用于设置有多少个子进程负责加载数据。其工作原理为主进程启动 num_worker 个子进程（worker），每个进程所需采样点的索引，由主进程负责产生并分配。然后主进程依次轮询各个 worker，获得其各自生成的 batch。因此这种多进程读取方式可以在 GPU 进行密集运算时同步处理数据读取，抵消掉处理 io 的延时。num_worker 的设置并不建议越大越好，因为过多的子进程会占据 CPU 计算资源，使得程序中其他在 CPU 上的计算部分变慢，导致整体运行时间增加。一般来说是通过逐步增加尝试来进行设置。比如当 GPU 计算利用率已经很饱和时，说明数据读取足够满足计算需求，则不必再增加 worker 数量。

然后是 collate_fn 的设置。默认情况下 DataLoader 将调用预置的 default_collate_fn，将 Dataset 的返回的多个数据样本整理（collate）成为一个 batch。在 collate 时，会添加一个维度，即批样本维度在数据的第一维。可以看做这个操作即是 torch.stack 运算。在一般图像处理领域不涉及变长度样本情况下，使用 default_collate_fn 即可。但是当面对如文本或多标签分类问题需要处理变长样本特征问题时，就需要自己实现 collate_fn。这里我们用一个模拟例子展示其用法。

```python
import torch
from torch.utils import data

class FakeData(data.Dataset):
    def __init__(self, max_len):
        self.max_len = max_len

    def __getitem__(self, item):            # 根据索引返回[0, item]之间的整数
        return torch.arange(item+1)

    def __len__(self):                      # 最多只有max_len个数据点
        return self.max_len


def custom_collate_fn(batch):
    lengths = [len(b) for b in batch]        # 获取每个数据点长度
    data = torch.zeros(len(batch), max(lengths))   # 确定最大数据维度
    for i in range(len(batch)):
        end = lengths[i]
        data[i, :end] = batch[i]             # 0:end填原数据，剩下补零
```

```
    return data

loader = data.DataLoader(
    FakeData(10), batch_size=5,
    shuffle=True, collate_fn=custom_collate_fn
)

for data in loader:
    print(data)
```

以上运行结果为：
```
tensor([[0., 1., 2., 3., 4., 5., 0., 0., 0.],
        [0., 1., 2., 3., 4., 5., 6., 7., 8.],
        [0., 1., 2., 3., 4., 0., 0., 0., 0.],
        [0., 1., 2., 3., 4., 5., 6., 7., 0.],
        [0., 1., 2., 3., 0., 0., 0., 0., 0.]])
tensor([[0., 1., 2., 3., 4., 5., 6., 0., 0., 0.],
        [0., 1., 2., 0., 0., 0., 0., 0., 0., 0.],
        [0., 0., 0., 0., 0., 0., 0., 0., 0., 0.],
        [0., 1., 0., 0., 0., 0., 0., 0., 0., 0.],
        [0., 1., 2., 3., 4., 5., 6., 7., 8., 9.]])
```

可以看出一个完整迭代 DataLoader 的过程是将所有数据输出，并且按照每次 batch 中最长样本点进行补零。这对于处理变长数据格式时很有帮助。

最后就是 pin_memory 的设置。由于从 CPU 数据转移至 GPU 时，位于 pinned（或叫做 page-locked）memory 上的 Tensor 会更快，因此 DataLoader 里设置了这一选项，如果设置为 pin_memory=True，则在提供数据时，调用 Tensor 的.pin_memory()方法提高转移速度。但是该方法只对普通 Tensor 和包含 Tensor 的映射与容器等数据结构成立，如果是自定义的数据 batch，则需要特殊实现其.pin_memory 方法。

以上便是从使用角度介绍了 PyTorch 中数据加载的机制，对于其内部实现机制可以参考其源码文档讲解[①]。另外当使用分布式计算时，数据流处理方式还有不同，这一部分内容将在第 5 章中介绍。

## 4.2　PyTorch 模型搭建

在第 3 章中我们已经介绍了利用 nn.Module 搭建深度神经网络模型。这套流程对于重新设计并自己实现网络是比较自由方便的，适用于一般小型的任务。但是在实际应用当中，我们往往会使用学术界广为认同的经典模型作为实验基础。因此这涉及到如何复用经典模型，如何加载预训练模型，如何跨框架跨语言分享模型等情况。在本节中我们将对此一一进行介绍。

---

① https://pytorch.org/docs/stable/_modules/torch/utils/data/dataloader.html#DataLoader

## 4.2.1 经典模型复用与分享：torchvision.models

在 torchvision.models 模块中，包含了在计算机视觉中大量的经典模型。不仅包括了标准的实现代码，同时也提供了预训练权重。这为后续的诸多研究工作提供了方便。其使用方式非常简单：

```
>> from torchvision import models
>> net = models.resnet50()
>> net = models.mobilenet_v1()
>> net = models.resnet50(pretrained=True)
>> net = models.mobilenet_v1(pretrained=True)
```

其中当设置 pretrained=True 时，则在搭建好模型之后，加载预训练权重作为初始化。如果该权重之前没有下载过，将先花费一段时间下载，并存储在本地 ~/.cache/torch/checkpoints 文件夹中，以便于下次直接加载。这些分类模型都是在 ImageNet 数据集上进行训练，输入图片均按照相同的归一化进行处理：

```
normalize = transforms.Normalize(mean=[0.485, 0.456, 0.406],
                                 std=[0.229, 0.224, 0.225])
```

因此省却了很多不同模型预处理带来的麻烦。这些模型在训练或测试时需要设置其为 .train() 或 .eval()，因为其中往往包含 nn.BatchNorm 和 nn.Dropout 模块，如果忘记设置将会引起预测失效。torchvision.models 中所有实现的分类模型总结在表 4.3 中。

表 4.3 torchvision.models 中所有实现的分类模型

| AlexNet | VGG-13-bn | ResNet-101 | Densenet-201 | ResNeXt-50-32x4d |
| VGG-11 | VGG-16-bn | ResNet-152 | Densenet-161 | ResNeXt-101-32x8d |
| VGG-13 | VGG-19-bn | SqueezeNet | Inception-V3 | Wide ResNet-50-2 |
| VGG-16 | ResNet-18 | SqueezeNet | GoogleNet | Wide ResNet-101-2 |
| VGG-19 | ResNet-34 | Densenet-121 | ShuffleNet-V2 | MNASNet 1.0 |
| VGG-11-bn | ResNet-50 | Densenet-169 | MobileNet-V2 | |

然而毕竟计算机视觉领域发展突飞猛进，各种各样的模型层出不穷，而且最近神经架构搜索（Neural Architecture Search，NAS）研究领域的火热，使得新型模型架构的设计和发现更加容易且多样。因此一些第三方代码库则迅速跟进，实现了最新的模型。对于计算机视觉分类模型来说，推荐 pytorch-image-models[①] 这个代码库，这里面包括了诸多最新的模型实现，如 EfficientNet[37]、MixNet[38]、MobileNet-V3[39] 等等。

在新版本的 PyTorch 中，添加了 torch.hub 作为一种新的预训练模型分享方法。它可以方便研究者在自己的 github 代码仓库中添加 hubconf.py 配置文件，上传自己的预训练

---

① https://github.com/rwightman/pytorch-image-models

模型配置。而其他用户可以 torch.hub.load 方式简单使用。但目前 torch.hub 还处在开发阶段，使用方法和 API 接口也会改变，但可以保持后续关注。

torchvision.models 模块也在更新后增加了语义分割、物体检测、实例分割、关键点检测和视频分类模型，进一步扩充了计算机视觉中常用任务中的预训练模型。但是当前这些模型还不容易改造成为可训练模式，因为这些任务本身涉及的监督信号和数据增广流程较为复杂。但是可以在测试模式下作为特征提取的应用，对于快速构建整体应用原型非常有帮助。

### 4.2.2 模型加载与保存

在上一节中介绍了可以通过 torchvision.models 一键加载预训练模型。如果深入其中的实现方法，可以看到其实就是简单的两步，具体实现为：

```
model = torchvision.models.SomeModel()
model.load_state_dict(torch.load('pretrained_weights.pth'))
```

其中 torch.load 读取预训练模型权重到内存，model.load_state_dict 则是根据名称一一对应赋值模型内的权重。这里预训练权重是一个字典，包含了模型中参数名称与相应数值。而这个预训练权重字典也是很容易获得，只需要如下实现：

```
torch.save(model.state_dict(), 'model_weights.pth')
```

其中 model.state_dict 获取得到模型中所有待存储参数，torch.save 则是将其持久化存储。这里的参数不仅包括各计算层可训练参数，也包括 nn.Module 中的缓冲区变量（buffer parameter），比如 nn.BatchNorm 层中统计的均值和标准差。如果原模型运行在 GPU 而想要保存在 CPU 上，需要先 model.to('cpu') 再保存。

以上可以看出 PyTorch 中模型保存时是将权重张量与计算图分离，因此仅获取权重是无法重构出计算图的。虽然这种做法带来部署的麻烦，但是也方便了不同框架之间的模型转换。因为只需要在 PyTorch 中构建出相同的计算图，并将另外一种框架的预训练权重按照权重名称和对应值的字典存储好，就可以在 PyTorch 中加载。这里我们展示一个例子，在 PyTorch 中构建 GoogLeNet 模型，并加载由 Caffe 框架得到的预训练权重。

首先我们可以先了解 GoogLeNet 的模型架构，并观察 Caffe 中的命名规则。这样在导入权重时就很方便对应。这里推荐一个可视化 Caffe 模型配置文件的网站[①]。图 4.4 中展示了 GoogLeNet 模型局部可视化情况，右侧表格则详细展示了其架构组成。

---

① http://ethereon.github.io/netscope/#/preset/googlenet

| type | patch size/ stride | output size | depth |
| --- | --- | --- | --- |
| convolution | 7×7/2 | 112×112×64 | 1 |
| max pool | 3×3/2 | 56×56×64 | 0 |
| convolution | 3×3/1 | 56×56×192 | 2 |
| max pool | 3×3/2 | 28×28×192 | 0 |
| inception (3a) | | 28×28×256 | 2 |
| inception (3b) | | 28×28×480 | 2 |
| max pool | 3×3/2 | 14×14×480 | 0 |
| inception (4a) | | 14×14×512 | 2 |
| inception (4b) | | 14×14×512 | 2 |
| inception (4c) | | 14×14×512 | 2 |
| inception (4d) | | 14×14×528 | 2 |
| inception (4e) | | 14×14×832 | 2 |
| max pool | 3×3/2 | 7×7×832 | 0 |
| inception (5a) | | 7×7×832 | 2 |
| inception (5b) | | 7×7×1024 | 2 |
| avg pool | 7×7/1 | 1×1×1024 | 0 |
| dropout (40%) | | 1×1×1024 | 0 |
| linear | | 1×1×1000 | 1 |
| softmax | | 1×1×1000 | 0 |

图 4.4

可以看出其主要组成为反复堆叠的 Inception 模块，该模块内部由多路分支组成，其内部示意图如图 4.5 所示（取自原论文）。其中分为 4 路，采用不同的卷积和池化对输入特征进行转换，并最终拼接在一起形成输出。

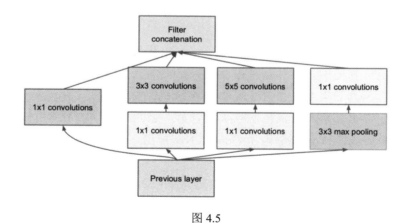

图 4.5

因此我们可以在 PyTorch 中实现的 Inception 模块如下：

```
import torch.nn as nn
import torch
```

```python
class Inception(nn.Module):
    def __init__(self, in_channel, br_1x1,
            br_3x3_reduce, br_3x3,
            br_5x5_reduce, br_5x5, pool_proj):
        super(Inception, self).__init__()
        self.add_module(
            '1x1', nn.Conv2d(in_channel, br_1x1, kernel_size=1)
        )
        self.add_module('relu_1x1', nn.ReLU())

        self.add_module(
            '3x3_reduce',
            nn.Conv2d(in_channel, br_3x3_reduce, kernel_size=1)
        )
        self.add_module('relu_3x3_reduce', nn.ReLU())
        self.add_module(
            '3x3',
            nn.Conv2d(br_3x3_reduce, br_3x3, kernel_size=3, padding=1)
        )
        self.add_module('relu_3x3', nn.ReLU())

        self.add_module(
            '5x5_reduce',
            nn.Conv2d(in_channel, br_5x5_reduce, kernel_size=1)
        )
        self.add_module('relu_5x5_reduce', nn.ReLU())
        self.add_module(
            '5x5',
            nn.Conv2d(br_5x5_reduce, br_5x5, kernel_size=5, padding=2)
        )
        self.add_module('relu_5x5', nn.ReLU())

        self.add_module(
            'pool',
            nn.MaxPool2d(kernel_size=3, stride=1, padding=1)
        )
        self.add_module(
            'pool_proj',
            nn.Conv2d(in_channel, pool_proj, kernel_size=1)
        )
        self.add_module('relu_pool_proj', nn.ReLU())

    def forward(self, x):
        x1 = getattr(self, '1x1')(x)
        x1 = getattr(self, 'relu_1x1')(x1)

        x2 = getattr(self, '3x3_reduce')(x)
        x2 = getattr(self, 'relu_3x3_reduce')(x2)
```

```
            x2 = getattr(self, '3x3')(x2)
            x2 = getattr(self, 'relu_3x3')(x2)

            x3 = getattr(self, '5x5_reduce')(x)
            x3 = getattr(self, 'relu_5x5_reduce')(x3)
            x3 = getattr(self, '5x5')(x3)
            x3 = getattr(self, 'relu_5x5')(x3)

            x4 = getattr(self, 'pool')(x)
            x4 = getattr(self, 'pool_proj')(x4)
            x4 = getattr(self, 'relu_pool_proj')(x4)

            return torch.cat([x1, x2, x3, x4], dim=1)
```

这里的命名方式与 Caffe 中的实现保持一致，方便之后权重加载。接下来我们就可以利用 Inception block 搭建整体的 GoogLeNet 模型。这里再注意一点，原始 GoogLeNet 模型在训练时会增加额外中间层的分支，用来增强监督信息。但是在测试时，只保留最后层的输出。此处我们只实现测试情形时的网络。

```
class GoogLeNet(nn.Module):
    def __init__(self):
        super(GoogLeNet, self).__init__()
        self.add_module(
            'conv1/7x7_s2',
            nn.Conv2d(3, 64, kernel_size=7, stride=2, padding=3)
        )
        self.add_module('conv1/relu_7x7', nn.ReLU())

        self.add_module(
            'pool1', nn.MaxPool2d(3, stride=2, ceil_mode=True)
        )

        self.add_module('norm1', nn.LocalResponseNorm(5))

        self.add_module(
            'conv2/3x3_reduce', nn.Conv2d(64, 64, kernel_size=1)
        )
        self.add_module('conv2/relu_3x3_reduce', nn.ReLU())
        self.add_module(
            'conv2/3x3',
            nn.Conv2d(64, 192, kernel_size=3, padding=1)
        )
        self.add_module('conv2/relu_3x3', nn.ReLU())

        self.add_module('norm2', nn.LocalResponseNorm(5))

        self.add_module(
            'pool2', nn.MaxPool2d(3, stride=2, ceil_mode=True)
```

```python
)
self.inception_3a = Inception(
    192, 64, 96, 128, 16, 32, 32
)
self.inception_3b = Inception(
    256, 128, 128, 192, 32, 96, 64
)

self.add_module(
    'pool3', nn.MaxPool2d(3, stride=2, ceil_mode=True)
)

self.inception_4a = Inception(
    480, 192, 96, 208, 16, 48, 64
)
self.inception_4b = Inception(
    512, 160, 112, 224, 24, 64, 64
)
self.inception_4c = Inception(
    512, 128, 128, 256, 24, 64, 64
)
self.inception_4d = Inception(
    512, 112, 144, 288, 32, 64, 64
)
self.inception_4e = Inception(
    528, 256, 160, 320, 32, 128, 128
)

self.add_module(
    'pool4', nn.MaxPool2d(3, stride=2, ceil_mode=True)
)

self.inception_5a = Inception(
    832, 256, 160, 320, 32, 128, 128
)
self.inception_5b = Inception(
    832, 384, 192, 384, 48, 128, 128
)

self.add_module('pool5', nn.AvgPool2d(7, stride=1))

self.add_module('drop', nn.Dropout2d(p=0.4))

self.add_module(
    'loss3/classifier', nn.Linear(1024, 1000)
)
```

```python
def forward(self, x):
    x = getattr(self, 'conv1/7x7_s2')(x)
    x = getattr(self, 'conv1/relu_7x7')(x)
    x = self.pool1(x)
    x = self.norm1(x)

    x = getattr(self, 'conv2/3x3_reduce')(x)
    x = getattr(self, 'conv2/relu_3x3_reduce')(x)
    x = getattr(self, 'conv2/3x3')(x)
    x = getattr(self, 'conv2/relu_3x3')(x)
    x = self.norm2(x)
    x = self.pool2(x)

    x = self.inception_3a(x)
    x = self.inception_3b(x)
    x = self.pool3(x)

    x = self.inception_4a(x)
    x = self.inception_4b(x)
    x = self.inception_4c(x)
    x = self.inception_4d(x)
    x = self.inception_4e(x)
    x = self.pool4(x)

    x = self.inception_5a(x)
    x = self.inception_5b(x)
    x = self.pool5(x)
    x = self.drop(x)

    x = x.view(-1, 1024)
    x = getattr(self, 'loss3/classifier')(x)

    return
```

接下来就是如何转换 Caffe 的预训练权重，并加载到 PyTorch 模型中。首先，我们需要从公开链接[1]下载开源的预训练权重。然后，为了可以解析出其中的内容，需要在 Caffe 代码仓库链接[2]下载其 Protobuf 格式文件 caffe.proto。这里我们采取了最简单的方法去解析权重内容，而没有从头开始安装 Caffe。然后运行如下命令：

```
>> protoc --python_out=. caffe.proto
```

生成出 caffe_pb2.py 文件。利用该文件就可以实现在 PyTorch 加载并解析 Caffe 预训练权重了。接下来利用如下程序按照对应权重名称加载预训练模型，并保存为 PyTorch 模型。

---

[1] http://dl.caffe.berkeleyvision.org/bvlc_googlenet.caffemodel

[2] https://raw.githubusercontent.com/BVLC/caffe/master/src/caffe/proto/caffe.proto

```python
import caffe_pb2
import numpy as np
from model import GoogLeNet
import torch

with open('bvlc_googlenet.caffemodel', 'rb') as f:
    net = caffe_pb2.NetParameter()
    net.ParseFromString(f.read())              # 解析 Caffe 模型权重

net_params = {}

for i in range(len(net.layers)):               # 存储为字典，方便 PyTorch 模型加载
    if len(net.layers[i].blobs) > 0:
        net_params[net.layers[i].name + '/weight'] = \
            np.array(net.layers[i].blobs[0].data, dtype=np.float32)
        net_params[net.layers[i].name + '/bias'] = \
            np.array(net.layers[i].blobs[1].data, dtype=np.float32)

net = GoogLeNet()
for name, p in net.named_parameters():
    name = name.split('.')                      # 按照名称一一对应加载模型
    caffe_name = '/'.join(name)
    p.data = torch.from_numpy(
        net_params[caffe_name].reshape(p.size())
    )

torch.save(net.state_dict(), 'googlenet.pth')  # 保存模型
```

图 4.6 展示了在给定左侧输入图片后，PyTorch 实现的 GoogLeNet 模型给出的 top-5 预测概率及对应类别名称。可以看出成功实现了准确预测，说明模型转换成功。

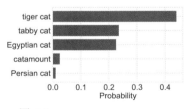

图 4.6

### 4.2.3 导出为 ONNX 格式

上一节展示了如何手动将 Caffe 预训练权重按照字典方式加载到 PyTorch 中。但是这样的过程过于麻烦且容易出错。如何能在多种框架之间进行模型转换共享是一个需要统一解决的问题。这里我们介绍开放神经网络交换（Open Neural Network Exchange，ONNX）格式。这是由微软主导开发，用来统一表征神经网络的一种格式。它可以将某一种框架

中的模型导出为 ONNX 中间表示（intermediate representation，IR），然后在另一种框架中用对应的算子重构出模型，实现了不同框架之间模型的转换与分享。当前诸多框架已经支持 ONNX 格式导出，如 TensorFlow、PaddlePaddle、MXNet、Chainer、CoreML、CNTK 等等。在 PyTorch 中，模型导出为 ONNX 的方式也很简单。此处我们以导出 AlexNet 模型为例：

```
>> import torchvision.models as models
>> import torch
>> alexnet = models.alexnet(pretrained=True)
>> alexnet.eval()
>> dummy_input = torch.randn(1, 3, 224, 224)
>> torch.onnx.export(alexnet, dummy_input, 'alexnet.onnx', verbose=True)
```

最后一行通过指定输入导出 ONNX 格式的模型 alexnet.onnx，其中包含了模型结构与参数。verbose=True 则会输出如下可读的网络结构信息：

```
graph(%input.1 : Float(1, 3, 224, 224),
      %features.0.weight : Float(64, 3, 11, 11),
      %features.0.bias : Float(64),
      %features.3.weight : Float(192, 64, 5, 5),
      %features.3.bias : Float(192),
      %features.6.weight : Float(384, 192, 3, 3),
      %features.6.bias : Float(384),
      %features.8.weight : Float(256, 384, 3, 3),
      %features.8.bias : Float(256),
      %features.10.weight : Float(256, 256, 3, 3),
      %features.10.bias : Float(256),
      %classifier.1.weight : Float(4096, 9216),
      %classifier.1.bias : Float(4096),
      %classifier.4.weight : Float(4096, 4096),
      %classifier.4.bias : Float(4096),
      %classifier.6.weight : Float(1000, 4096),
      %classifier.6.bias : Float(1000)):
  %17 : Float(1, 64, 55, 55) = onnx::Conv[dilations=[1, 1], group=1,
      kernel_shape=[11, 11], pads=[2, 2, 2, 2], strides=[4,
      4]](%input.1, %features.0.weight, %features.0.bias)
  %18 : Float(1, 64, 55, 55) = onnx::Relu(%17)
  %19 : Float(1, 64, 27, 27) = onnx::MaxPool[kernel_shape=[3, 3], pads=[0,
      0, 0, 0], strides=[2, 2]](%18)
  ...
  ...
  %41 : Float(1, 4096) = onnx::Relu(%40)
  %42 : Float(1, 1000) = onnx::Gemm[alpha=1, beta=1,
      transB=1](%41, %classifier.6.weight, %classifier.6.bias)
  return (%42)
```

然后我们可以加载 alexnet.onnx 观察其结果。这一步骤需要事先 pip install onnx，然后运行以下命令：

```
>> import onnx
>> model = onnx.load_model('alexnet.onnx')
>> onnx.checker.check_model(model)              # 检查 IR 是否完整合理
>> print(onnx.helper.printable_graph(model.graph))   # 输出模型计算图
graph torch-jit-export (
  %input.1[FLOAT, 1x3x224x224]
) initializers (
  %classifier.1.bias[FLOAT, 4096]
  %classifier.1.weight[FLOAT, 4096x9216]
  %classifier.4.bias[FLOAT, 4096]
  %classifier.4.weight[FLOAT, 4096x4096]
  %classifier.6.bias[FLOAT, 1000]
  %classifier.6.weight[FLOAT, 1000x4096]
  %features.0.bias[FLOAT, 64]
  %features.0.weight[FLOAT, 64x3x11x11]
  %features.10.bias[FLOAT, 256]
  %features.10.weight[FLOAT, 256x256x3x3]
  %features.3.bias[FLOAT, 192]
  %features.3.weight[FLOAT, 192x64x5x5]
  %features.6.bias[FLOAT, 384]
  %features.6.weight[FLOAT, 384x192x3x3]
  %features.8.bias[FLOAT, 256]
  %features.8.weight[FLOAT, 256x384x3x3]
) {
  %17 = Conv[dilations = [1, 1], group = 1, kernel_shape = [11, 11],
    pads = [2, 2, 2, 2], strides = [4, 4]]
    (%input.1, %features.0.weight, %features.0.bias)
  %18 = Relu(%17)
  %19 = MaxPool[kernel_shape = [3, 3], pads = [0, 0, 0, 0], strides = [2, 2]](%18)
  %20 = Conv[dilations = [1, 1], group = 1, kernel_shape = [5, 5], pads
= [2, 2, 2, 2], strides = [1, 1]](%19, %features.3.weight, %features.3.bias)
  %21 = Relu(%20)
  %22 = MaxPool[kernel_shape = [3, 3], pads = [0, 0, 0, 0], strides = [2, 2]](%21)
  ...
  ...
```

有了 ONNX 格式的模型，可以脱离 PyTorch 框架做很多应用。首先其可以方便展示模型架构。之前有介绍展示 Caffe 模型架构的链接。这里介绍另外一个链接[①]，可以展示 ONNX 模型，因此可以展示多种框架的模型结构。像之前导出的 alexnet.onnx 在上传到该网站后可视化结果如图 4.7 所示。其中包含了非常详细的属性描述。

---

① https://lutzroeder.github.io/netron/

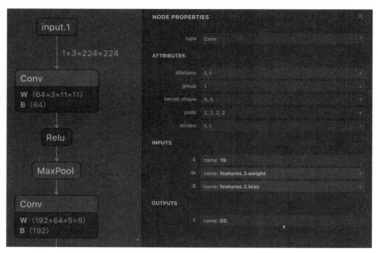

图 4.7

而且 ONNX 模型格式可以在不同框架间进行转换。例如 PyTorch 官方教程[①]中介绍了如何将一个风格迁移的 PyTorch 模型导出到 ONNX 格式，再从 CoreML 框架中加载，并最终部署到 iOS 应用中。ONNX 本身也支持跨平台跨设备运行，并且包含了结构优化，因此以后可以用易用性更好的动态框架进行训练，然后通过 ONNX 格式导出，并部署在移动端设备实现高速推理。

利用 ONNX 格式还可以自由地在其他语言中进行扩展。微软官方代码库[②]展示了在 JavaScript 语言中支持 ONNX 模型运行，从而可以在网页中嵌入类似如下的代码，实现在本地浏览器中运行神经网络的效果。更多例子可参考 online demo 网站[③]。

```
async function predict() {
  const sess = new onnx.InferenceSession({backendHint: 'webgl'});
  await sess.loadModel('./alexnet.onnx');
  const imageLoader = new ImageLoader(224, 224);
  const imageData = await imageLoader.getImageData('./input.jpg');
  const preprocessedData = preprocess(imageData.data, 224, 224);
  const input = new onnx.Tensor(
    preprocessedData, 'float32', [1, 3, 224, 224]
  );
  const outputMap = await sess.run([input]);
  const outputTensor = outputMap.values().next().value;
  const prediction = outputTensor.data;
  const i = prediction.indexOf(Math.max(...prediction));
  console.log('Output Class Name: ${className[i]},
               Probability: ${prediction[i]}'
```

---

① https://pytorch.org/tutorials/advanced/ONNXLive.html?highlight=onnx
② https://github.com/Microsoft/onnxjs
③ https://microsoft.github.io/onnxjs-demo/#/

```
    );
}
```

## 4.3 训练过程中日志记录与可视化

在训练模型过程中，除了以上的数据处理和模型搭建外，还需要实时追踪训练过程中的结果。这就涉及到训练日志记录，以及一些中间结果可视化。在之前的实战环节中，我们只是将训练和测试过程中的损失函数值以及准确率等信息直接输出到终端，虽然简单直观，但是既不利于保存日志，也无法看出指标的直观趋势。在本节中我们将介绍，在 PyTorch 中如何利用 tensorboard 实现以上功能。

tensorboard 原是属于 TensorFlow 框架中的一个日志记录和可视化工具。但是因其使用直观方便，受到其他深度学习框架社区的欢迎。在 PyTorch 官方集成对于 tensorboard 的支持之前，就已经有一些第三方实现如 tensorboardX 通过改造封装，可以实现在 tensorboard 中显示 PyTorch 中的张量信息。而此次我们则选择官方实现 torch.utils.tensorboard 模块。

在使用 tensorboard 之前，首先要安装 TensorFlow 和 tensorboard。这两者均可以通过 pip install 实现安装，建议安装版本号至少为 1.15。此处介绍在 PyTorch 中导入 tensorboard。

```
from torch.utils.tensorboard import SummaryWriter

writer = SummaryWriter('tb-logs')
...
...
writer.close()
```

这里我们初始化了一个日志记录写入器，并指定其写入日志的文件夹为'tb-logs'。中间为记录日志操作，最后结尾调用 writer.close()结束。

首先，最简单的记录数值方法 writer.add_scalar。

```
import torch

for i in torch.arange(100):
    writer.add_scalar('Loss/sin_x', torch.sin(i/5.0), i)
    writer.add_scalar('Loss/cos_x', torch.cos(i/5.0), i)
```

其参数列表为(tag, value, global_step)，分别指定该值的标签名称、数值大小和全局步数值。其中'MetaTag/Name'形式的标签表示所属大类别和具体的名称。同一个 MetaTag 的多个图将会总结在一起。上面这种添加两个值的写法，也可以用 writer.add_scalars 方法同时添加到一个图中：

```
for i in torch.arange(100):
    writer.add_scalars('Loss/two_losses',
                       {'sin_x': torch.sin(i/5.0),
                        'cos_x': torch.cos(i/5.0)}, i)
```

这时所接受的不是单个数值，而是字典标明每个值的名称。此时在终端运行如下命令启动 tensorboard 运行文件夹'tb-logs'中记录的日志：

```
$ tensorboard --logdir=tb-logs
TensorBoard 1.15.0 at http://local:6006/ (Press CTRL+C to quit)
```

这时在本地浏览器中打开 localhost:6006 链接则会显示以上绘图结果，如图 4.8 所示。

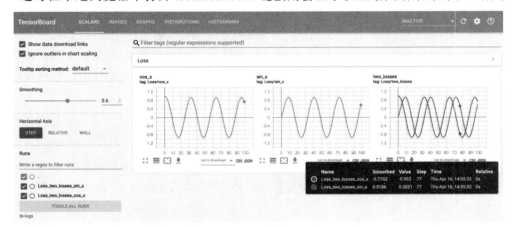

图 4.8

在训练过程中，除了记录数值之外，有时我们还希望观察如权重等张量总体数值分布情况。此时可以使用 writer.add_histogram 方法实现，具体使用方法与 writer.add_scalar 类似。

```
for i in torch.arange(10):
    mean = torch.sin(i/2.0)
    random_weights = torch.normal(mean, 0.1, (100,))
    writer.add_histogram('Weights/w', random_weights, i)
```

这是重新载入 tensorboard 页面，并选择 HISTOGRAMS 选项卡，则可以看到图 4.9 显示的结果。

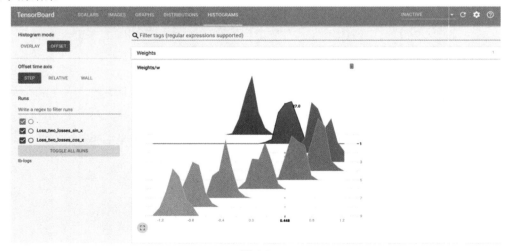

图 4.9

tensorboard 除了以上显示数值等方法以外，还可以展示图片。比如在训练之初显示由 dataloader 产生的数据，可以初步检查数据加载格式是否正确。这里使用 writer.add_images 方法，用于批量显示多张图片。注意此处想要正确显示图片，需要保证输入范围在[0, 1]。因此这里只使用了 transforms.ToTensor 变换。writer.add_images 接受的输入形状为 NCHW，与 dataloader 产生的批量数据形状一致，因此无需变换形状。

```
from torchvision import datasets, transforms

train_loader = torch.utils.data.DataLoader(
    datasets.MNIST(
        './data', train=True, download=False,
        transform=transforms.Compose([
            transforms.ToTensor(),
        ])
    ),
    batch_size=64, shuffle=True)

train_loader = iter(train_loader)
batch, label = next(train_loader)
writer.add_images('Images/mnist', batch, 0)
```

然后重新加载 tensorboard 界面，在 IMAGES 选项卡下就可以看到图 4.10 所示结果。可以看到其正常展示了 MNIST 数字图片。

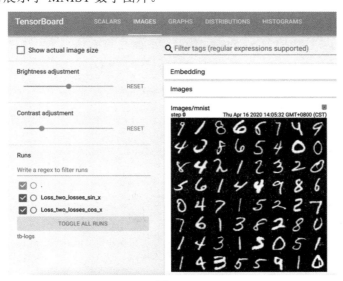

图 4.10

而如果有些时候，需要进行较为复杂的画图。这时可以使用 writer.add_figure 方法进行展示。只需要先按照正常 matplotlib[①]画图方法进行，然后将该图的手柄（handle）传入

---

① https://matplotlib.org

writer.add_figure 方法即可。此处我们展示一个将 MNIST 数字图片转化为 784 维向量，并使用 t-SNE[40]降维的方法，在二维平面上画出其投影嵌入表示，并在 tensorboard 中呈现的流程。在实现之前需要 pip install scikit-learn matplotlib。

```
from sklearn.manifold import TSNE
import matplotlib.pyplot as plt
import matplotlib

batch = batch.view(64, 28*28).numpy()  # batch和label为之前dataloader提供
label = label.numpy()
embedding = TSNE(2).fit_transform(batch)    # 进行降维产生二维投影嵌入

fig = plt.figure(figsize=(5, 5))
cmap = plt.cm.hsv
norm = matplotlib.colors.BoundaryNorm(np.arange(0, 12), cmap.N)
plt.scatter(embedding[:, 0], embedding[:, 1], s=10,      # 绘制嵌入散点图
            c=label, edgecolors='k', linewidths=1, alpha=1,
            cmap=cmap, norm=norm)

writer.add_figure('Embedding/mnist', fig, 0) # 在tensorboard中展示
```

然后重新加载 tensorboard 界面，在 IMAGES 选项卡下就可以看到图 4.11 所示结果。其实针对展示图片特征二维嵌入的画图，也可以直接使用 tensorboard 中内置的方法 writer.add_embedding，读者可以自己探索。

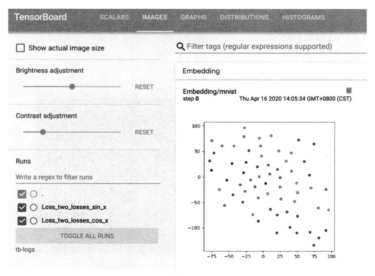

图 4.11

最后 tensorboard 还提供了可视化 PyTorch 模型网络结构的方法 writer.add_graph。其参数列表为(model, input_to_model, verbose)。这里我们展示可视化 torchvision 中已实现的 AlexNet 网络结构：

```
from torchvision import models

model = models.alexnet()
model.eval()
input = torch.randn(1, 3, 224, 224)
writer.add_graph(model, input_to_model=input)
```

然后重新加载 tensorboard 界面，在 GRAPHS 选项卡下就可以看到图 4.12 所示结果。其中每一个高层模块还可以进行放大展开其内部结构，交互方式非常直观方便。

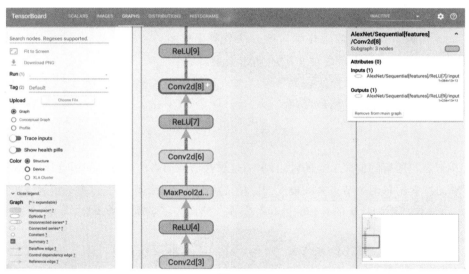

图 4.12

除了以上介绍的一些方法，tensorboard 还支持展示视频、文本、音频，添加精度召回曲线等等功能。更多使用方法及介绍可以参见 PyTorch 官方文档[①]，以及 TensorFlow 中关于 tensorboard 的介绍[②]。

## 4.4 PyTorch 应用实战一：在 CIFAR10 数据集进行神经网络结构搜索

随着近些年自动机器学习（AutoML）的兴起，神经网络结构搜索（Neural Architecture Search，NAS）课题逐渐成为当前深度学习领域研究的重点。NAS 相关技术和方法旨在通过算法自动对深度神经网络结构进行设计，从而筛选出超过一般手工设计网络的性能。在本节中，我们就以 CIFAR10 数据集为基础，在一个简化的网络空间中，应用可微分架构搜索方法 DARTS，实现神经网络结构搜索。

---

① https://pytorch.org/docs/stable/tensorboard.html

② https://www.tensorflow.org/tensorboard/

## 4.4.1 可微分网络架构搜索 DARTS 介绍

DARTS 方法来源于发表在 ICLR 2019 的文章"DARTS: Differentiable Architecture Search"[41]。其核心思想比较直观，将网络结构搜索问题转化成类似网络剪枝（Network Pruning）。首先将所有可待选的子模块都连接到网络中，然后模块直接的连接强度和权重均以端到端（End-to-End）可微分自动求导方式进行学习。然后根据连接强度去除掉作用较小的模块，从而实现从一个超网络（Hyper-Network）中挖掘出一个子网络。这个子网络的结构便被视为是 NAS 结果。最后再从头开始按照标准训练流程完整训练网络。由于这种方法无需像进化策略或强化学习等方法反复采样组合模块，并且以端到端可导方式训练，可以实现结构搜索阶段非常大的加速效果。

DARTS 方法中将每个模块视为节点。节点可以接受其他多个节点的输入，同时进行本模块操作并输出。这一过程可写作：

$$x^j = \sum_{i<j} o^{(i,j)}(x^i)$$

其中 $x^i$ 是 $x^j$ 所有的前序节点的输出，$o^{(i,j)}$ 是这些节点与当前节点之间所连接的运算操作。而为了学习搜索出每两个节点之间最优的操作，DARTS 在训练阶段先设置一组候选操作集合 $\mathcal{O}$，比如可能包括不同种类的卷积、池化等操作。然后为了实现可微求导，将 $o^{(i,j)}$ 建模为：

$$o^{(i,j)}(x) = \sum_{o \in \mathcal{O}} \frac{\exp\left(\alpha_o^{(i,j)}\right)}{\sum_{o' \in \mathcal{O}} \exp\left(\alpha_{o'}^{(i,j)}\right)} o(x)$$

即该运算操作的输出为所有候选操作输出的加权平均。这个权重系数由一组 $\alpha_o^{(i,j)}$ 经过 Softmax 函数转换而成。这样既满足加权系数和为 1，又可以可微地对系数求导，便于学习。该过程可以由图 4.13 展示（取自原论文）。其中（b）为完整的超网络，（c）为学习过后得到的权重系数，（d）则是根据权重系数筛选出的网络结构。

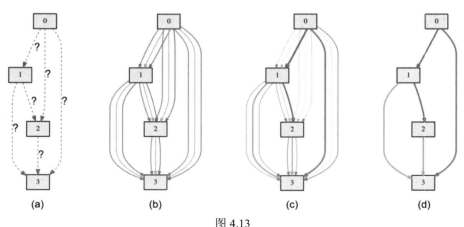

图 4.13

而如何训练整个超网络，原论文将此刻画为一个如下的 bilevel optimization 问题：

$$\min_{\alpha} \mathcal{L}_{val}(w^*(\alpha), \alpha)$$

$$s.t. \ w^*(\alpha) = \arg\min_{w} \mathcal{L}_{train}(w, \alpha)$$

其中 $w, \alpha$ 分别代表了模型的权重参数和结构参数。但是在实现时，完全解决 bilevel 优化问题计算复杂度过高，因此采取了近似做法，即采用交替方式，每次仅梯度下降后更新一步 $w$ 和 $\alpha$。

### 4.4.2 简化问题建模：以 ResNet 为例

这里为了便于展示，同时减少计算量。我们采取一个简化的网络模型，以 ResNet20 模型为例。原始的 ResNet20 模型结构中最重要的是残差模块（Residual Block），如图 4.14（a）所示，其中输入经过两层卷积之后与原始输入相加得到输出，3×3 代表卷积核大小，64 为卷积输出通道数目。而此处我们想要搜索的是中间两层卷积的类型，这里选取最简单候选运算集合，即包括 3×3、5×5 和 7×7 卷积核大小的三种卷积。因此改造后待搜索的残差模块如图 4.14（b）所示，输入进来之后经过三路不同卷积分支，最后利用加权系数求和，得到残差输出，并与原始输入相加。类似的想法也在论文[6]和[7]中使用。

最后我们模仿 ResNet20 原始网络结构，设计了超网络，如图 4.14（c）所示。由于此次实验在 CIFAR10 数据集上，因此输入为 $N \times 3 \times 32 \times 32$ 的张量。然后经过 3×3 的卷积、BatchNorm、ReLU 组合，输出通道数为 16。之后便是将图 4.13（b）中的待搜索模块 SearchBlock 进行反复堆叠，每个模块后数值代表该模块内部卷积输出通道数目，带有'/2'标识的模块要求第一次卷积的 stride 为 2，从而实现长宽降维。最后全局平均池化（GlobalAvgPool），输出 64 维特征向量，并经过最后 Linear 层得到 10 类别输出。

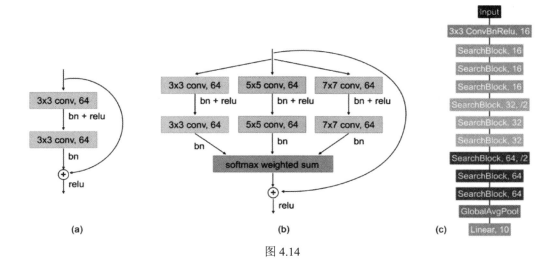

图 4.14

### 4.4.3 具体实现

接下来便是具体实现部分。首先我们实现待搜索模块及超网络，以下代码包括在 model_search.py 中。

```python
import torch
import torch.nn as nn
import torch.nn.functional as F


def residual_branch(in_channel,                      # 单个卷积核残差分支
                    mid_channel,                     # 第一层卷积跨度为 stride
                    out_channel,                     # 第二层卷积跨度为 1
                    kernel_size,
                    stride,
                    padding):
    return nn.Sequential(
        nn.Conv2d(in_channel, mid_channel,
                  kernel_size, stride,
                  padding, bias=False),
        nn.BatchNorm2d(mid_channel),
        nn.ReLU(),
        nn.Conv2d(mid_channel, out_channel,
                  kernel_size, 1,
                  padding, bias=False),
        nn.BatchNorm2d(out_channel),
    )

class BasicBlockSearch(nn.Module):                   # 待搜索模块
    def __init__(self,
                 in_channel,
                 mid_channels,
                 out_channel,
                 stride=1):
        super(BasicBlockSearch, self).__init__()
        Cin, Cout = in_channel, out_channel
        C3, C5, C7 = mid_channels
        self.mid_channels = mid_channels
        self.module_list = nn.ModuleList([           # 使用 ModuleList 表示三路分支
            residual_branch(Cin, C3, Cout, 3, stride, 1),
            residual_branch(Cin, C5, Cout, 5, stride, 2),
            residual_branch(Cin, C7, Cout, 7, stride, 3),
        ])
        self.relu = nn.ReLU()
        self.gates = nn.Parameter(torch.ones(3))     # 待学习的架构参数 α
        self.shortcut = nn.Sequential()              # 跨层连接
```

```python
            if stride != 1:                    # 当stride不为1时
                self.shortcut = nn.Sequential( # 使用1x1卷积降维
                    nn.Conv2d(Cin, Cout,
                              kernel_size=1,
                              stride=stride,
                              bias=False),
                    nn.BatchNorm2d(Cout)
                )

    def forward(self, x):
        residuals = []
        for m in self.module_list:              # 先计算各路分支输出
            residuals.append(m(x))

        probability = F.softmax(                # 经过softmax产生加权系数
            self.gates, dim=-1
        )
        merge_out = 0
        for r, p in zip(residuals, probability):# 加权求和
            merge_out += r * p

        out = self.shortcut(x) + merge_out
        out = self.relu(out)
        return out


class ResNetSearch(nn.Module):                  # 待搜索网络
    def __init__(self, depth=20, num_classes=10):
        super(ResNetSearch, self).__init__()
        n_blocks = (depth - 2) // 6             # 深度控制每个阶段模块重复次数

        self.in_channel = 16
        self.mid_channels = [16, 16, 16]
        self.conv = nn.Conv2d(
            3, self.in_channel,
            kernel_size=3, padding=1, bias=False
        )
        self.bn = nn.BatchNorm2d(self.in_channel)
        self.relu = nn.ReLU()
        self.layer1 = self._make_layer(n_blocks, 16, stride=1)
        self.layer2 = self._make_layer(n_blocks, 32, stride=2)
        self.layer3 = self._make_layer(n_blocks, 64, stride=2)
        self.avgpool = nn.AvgPool2d(8)
        self.linear = nn.Linear(64, num_classes)
        self._init_weights()

    # 搭建每阶段内部多个模块
```

```python
def _make_layer(self, n_blocks, out_channel, stride):
    multiplier = out_channel // self.in_channel
    self.mid_channels = [
        i * multiplier for i in self.mid_channels
    ]

    layers = []
    for i in range(n_blocks):
        layers.append(
            BasicBlockSearch(
                self.in_channel,
                self.mid_channels,
                out_channel,
                stride
            )
        )
        self.in_channel = out_channel
        stride = 1

    return nn.Sequential(*layers)

# 初始化权重
def _init_weights(self):
    for m in self.modules():
        if isinstance(m, nn.Conv2d):
            nn.init.kaiming_normal_(
                m.weight.data, nonlinearity='relu'
            )
        elif isinstance(m, nn.BatchNorm2d):
            m.weight.data.fill_(1)
            m.bias.data.zero_()

# 前向运算
def forward(self, x):
    x = self.conv(x)
    x = self.bn(x)
    x = self.relu(x)
    x = self.layer1(x)
    x = self.layer2(x)
    x = self.layer3(x)
    x = self.avgpool(x)
    x = x.view(-1, 64)
    x = self.linear(x)

    return x
```

接下来是一些辅助函数，包括保存日志文本、创建日志文件夹、加载和保存 pickle

文件等函数，总结在 misc.py 文件中。

```python
import os
import shutil
import pickle as pkl
import time
from datetime import datetime

# 日志
class Logger(object):
    def __init__(self):
        self._logger = None

    def init(self, logdir, name='log'):
        if self._logger is None:
            import logging
            if not os.path.exists(logdir):
                os.makedirs(logdir)
            log_file = os.path.join(logdir, name)
            if os.path.exists(log_file):
                os.remove(log_file)
            self._logger = logging.getLogger()
            self._logger.setLevel('INFO')
            fh = logging.FileHandler(log_file)
            ch = logging.StreamHandler()
            self._logger.addHandler(fh)
            self._logger.addHandler(ch)

    def info(self, str_info):
        now = datetime.now()
        display_now = str(now).split(' ')[1][:-3]
        self.init(os.path.expanduser('~/tmp_log'), 'tmp.log')
        self._logger.info('[' + display_now + ']' + ' ' + str_info)

logger = Logger()

# 创建文件夹
def ensure_dir(path, erase=False):
    if os.path.exists(path) and erase:
        print("Removing old folder {}".format(path))
        shutil.rmtree(path)
    if not os.path.exists(path):
        print("Creating folder {}".format(path))
        os.makedirs(path)

# 加载 pickle 文件
def load_pickle(path, verbose=True):
```

```python
    begin_st = time.time()
    with open(path, 'rb') as f:
        if verbose:
            print("Loading pickle object from {}".format(path))
        v = pkl.load(f)
    if verbose:
        print("=> Done ({:.4f} s)".format(time.time() - begin_st))
    return v

# 保存pickle文件
def dump_pickle(obj, path):
    with open(path, 'wb') as f:
        print("Dumping pickle object to {}".format(path))
        pkl.dump(obj, f, protocol=pkl.HIGHEST_PROTOCOL)

# 准备日志文件
def prepare_logging(args):
    args.logdir = os.path.join('./logs', args.logdir)

    logger.init(args.logdir, 'log')

    ensure_dir(args.logdir)
    logger.info("==================FLAGS==================")
    for k, v in args.__dict__.items():
        logger.info('{}: {}'.format(k, v))
    logger.info("=========================================")
```

然后便是训练 model_search，以实现可微网络结构搜索。以下代码包含在 train_search.py 中，其中一些超参数设置，参考 DARTS 公开的代码库[①]。这里由于搜索网络训练计算量较大，推荐使用 GPU 计算，默认 device='cuda'。在具体实现上，权重参数和架构参数使用两个不同的 optimizer，前者为 SGD，后者为 Adam。整体训练周期为 50，学习率调整器使用 CosineAnnealingLR。按照之前分解的 bilevel optimization 问题，需要区分出训练数据和验证数据，用以分别训练权重参数和架构参数。这里使用 50%的原始 CIFAR10 数据作为分割点，区分训练数据和验证数据。

```python
from model_search import ResNetSearch
from torchvision import transforms, datasets
from torch.utils.data.sampler import SubsetRandomSampler
from torch import optim
from torch.utils.tensorboard import SummaryWriter
import torch.nn as nn
import numpy as np
import argparse
import torch
```

---

① https://github.com/quark0/darts

```python
import os
import misc

print = misc.logger.info

parser = argparse.ArgumentParser()                    # 设置超参数
parser.add_argument('--gpu', default='0', type=str)
parser.add_argument('--depth', default=20, type=int)
parser.add_argument('--lr', default=0.01, type=float)
parser.add_argument('--lr_min', default=0.001, type=float)
parser.add_argument('--mm', default=0.9, type=float)
parser.add_argument('--wd', default=3e-4, type=float)
parser.add_argument('--epochs', default=50, type=int)
parser.add_argument('--batch_size', default=64, type=int)
parser.add_argument('--log_interval', default=100, type=int)
parser.add_argument('--train_portion', default=0.5, type=float)
parser.add_argument('--arch_lr', default=1e-3, type=float)
parser.add_argument('--arch_wd', default=1e-3, type=float)

args = parser.parse_args()                            # 解析超参数
args.num_classes = 10
args.device = 'cuda'
torch.backends.cudnn.benchmark = True                 # 启用 cudnn 加速
os.environ['CUDA_VISIBLE_DEVICES'] = args.gpu         # 设置目标 GPU 卡号

args.logdir = 'search-resnet%d' % args.depth          # 日志文件夹名

misc.prepare_logging(args)

print('==> Preparing data..')
transform = transforms.Compose([                      # 训练时数据预处理
    transforms.RandomCrop(32, padding=4),
    transforms.RandomHorizontalFlip(),
    transforms.ToTensor(),
    transforms.Normalize((0.4914, 0.4822, 0.4465),
                         (0.2023, 0.1994, 0.2010)),
])

train_data = datasets.CIFAR10(                        # 训练数据
    root='./data', train=True,
    download=True, transform=transform
)
num_train = len(train_data)
indices = list(range(num_train))
split = int(np.floor(args.train_portion * num_train))

train_loader = torch.utils.data.DataLoader(           # 利用 SubsetRandomSampler
```

```python
    train_data, batch_size=args.batch_size, # 使用前50%训练数据训练权重
    sampler=SubsetRandomSampler(indices[:split]),
    pin_memory=True, num_workers=2
)
val_loader = torch.utils.data.DataLoader(   # 使用后50%数据训练架构参数
    train_data, batch_size=args.batch_size,
    sampler=SubsetRandomSampler(indices[split:]),
    pin_memory=True, num_workers=2
)

print('==> Initializing model...')            # 初始化模型

model = ResNetSearch(args.depth, args.num_classes)
model.to(args.device)
criterion = nn.CrossEntropyLoss()
criterion.to(args.device)

model_params = []                             # 区分权重和架构参数
arch_params = []                              # 便于使用不同的优化器
for k, p in model.named_parameters():
    if k.endswith('gates'):
        arch_params.append(p)
    else:
        model_params.append(p)

arch_optim = optim.Adam(                      # 架构参数优化器
    arch_params, lr=args.arch_lr,
    betas=(0.5, 0.999), weight_decay=args.arch_wd
)
model_optim = optim.SGD(                      # 权重参数优化器
    model_params, lr=args.lr,
    momentum=args.mm, weight_decay=args.wd
)
scheduler = optim.lr_scheduler.CosineAnnealingLR( # 学习率调整器
    model_optim, args.epochs, eta_min=args.lr_min
)
writer = SummaryWriter(                       # tensorboard日志
    os.path.join(args.logdir, 'search-tb')
)

train_counter = 0
valid_counter = 0

def train(epoch):                             # 训练阶段
    model.train()
    for i, (data, target) in enumerate(train_loader):
        data = data.to(args.device)
```

```python
        target = target.to(args.device)

        data_arch, target_arch = next(iter(val_loader))
        data_arch = data_arch.to(args.device)
        target_arch = target_arch.to(args.device)

        arch_optim.zero_grad()                  # 先进行一步架构参数更新
        output = model(data_arch)
        loss_arch = criterion(output, target_arch)
        loss_arch.backward()
        arch_optim.step()

        model_optim.zero_grad()                 # 再进行一步权重参数更新
        output = model(data)
        loss_model = criterion(output, target)
        loss_model.backward()
        model_optim.step()

        if i % args.log_interval == 0:          # 输出训练信息
            acc = (output.max(1)[1] == target).float().mean()
            print('Train Epoch: %d [%d/%d] '
                  'Loss_A: %.4f, Loss_M: %.4f, Acc: %.4f' % (
                epoch, i, len(train_loader),
                loss_arch.item(), loss_model.item(), acc.item()
            ))
            global train_counter                # tensorboard写入信息
            writer.add_scalars('Loss/train_loss', {
                'loss_arch': loss_arch.item(),
                'loss_model': loss_model.item()
            }, train_counter)
            writer.add_scalar(
                'Accuracy/train_acc',
                acc.item(),
                train_counter
            )
            for k, p in enumerate(arch_params):  # 记录各层架构参数变化情况
                writer.add_scalars('Gates/gates_%d' % k, {
                    '3x3': p[0].item(),
                    '5x5': p[1].item(),
                    '7x7': p[2].item()
                }, train_counter)
            train_counter += 1


def evaluate(epoch):                             # 验证阶段
    model.eval()
    loss_avg = 0
```

```python
        acc_avg = 0
        with torch.no_grad():
            for i, (data, target) in enumerate(val_loader):
                data = data.to(args.device)
                target = target.to(args.device)

                output = model(data)
                loss = criterion(output, target)
                acc = (output.max(1)[1] == target).float().mean()
                loss_avg += loss.item()
                acc_avg += acc.item()

        loss_avg /= len(val_loader)                    # 计算平均 loss 和 acc
        acc_avg /= len(val_loader)
        print('...Evaluate @ Epoch: %d  Loss: %.4f, Acc: %.4f' % (
            epoch, loss_avg, acc_avg
        ))
        global valid_counter                           # tensorboard 写入信息
        writer.add_scalar('Loss/valid_loss', loss_avg, valid_counter)
        writer.add_scalar('Accuracy/valid_acc', acc_avg, valid_counter)
        valid_counter += 1


    for epoch in range(args.epochs):
        train(epoch)
        evaluate(epoch)
        scheduler.step()                               # 学习率调整
        torch.save(                                    # 模型权重保存
            model.state_dict(),
            os.path.join(args.logdir, 'model_search.pth')
        )
```

如果一切运行正常，则会在终端输出以下信息，并写入到 ./logs/search-resnet20/log 中。

```
$ python train_search.py --gpu 0 --depth 20
[15:43:44.931] =================FLAGS=================
[15:43:44.931] logdir: ./logs/search-resnet20
[15:43:44.931] log_interval: 100
[15:43:44.931] arch_lr: 0.001
[15:43:44.931] depth: 20
[15:43:44.931] lr_min: 0.001
[15:43:44.931] wd: 0.0003
[15:43:44.931] mm: 0.9
[15:43:44.931] arch_wd: 0.001
[15:43:44.931] gpu: 0
[15:43:44.931] epochs: 50
[15:43:44.932] batch_size: 64
[15:43:44.932] train_portion: 0.5
```

```
[15:43:44.932] num_classes: 10
[15:43:44.932] lr: 0.01
[15:43:44.932] device: cuda
[15:43:44.932] ========================================
[15:43:44.932] ==> Preparing data..
[15:43:45.924] ==> Initializing model...
[15:43:50.918] Train Epoch: 0 [0/391]    Loss_A: 2.3495, Loss_M: 2.2850, Acc: 0.1094
[15:44:24.307] Train Epoch: 0 [100/391]  Loss_A: 1.8521, Loss_M: 1.9227, Acc: 0.3438
[15:44:58.108] Train Epoch: 0 [200/391]  Loss_A: 1.7852, Loss_M: 1.6088, Acc: 0.3750
[15:45:32.071] Train Epoch: 0 [300/391]  Loss_A: 1.4048, Loss_M: 1.5785, Acc: 0.3750
[15:46:18.001] ...Evaluate @ Epoch: 0  Loss: 1.5446, Acc: 0.4352
[15:46:18.856] Train Epoch: 1 [0/391]    Loss_A: 1.5122, Loss_M: 1.4174, Acc: 0.5000
[15:47:06.939] Train Epoch: 1 [100/391]  Loss_A: 1.5036, Loss_M: 1.2741, Acc: 0.5312
[15:47:56.066] Train Epoch: 1 [200/391]  Loss_A: 1.4655, Loss_M: 1.3169, Acc: 0.4688
[15:48:44.356] Train Epoch: 1 [300/391]  Loss_A: 1.2794, Loss_M: 1.2198, Acc: 0.5469
[15:49:38.405] ...Evaluate @ Epoch: 1  Loss: 1.2919, Acc: 0.5276
```

同时也可以运行 tensorboard --logdir=./logs/search-resnet20/search-tb，在浏览器中打开 tensorboard 页面可以观察到如图 4.15 和图 4.16 所示结果，分别展示了训练过程中的损失函数、准确率和各层架构参数变化情况。

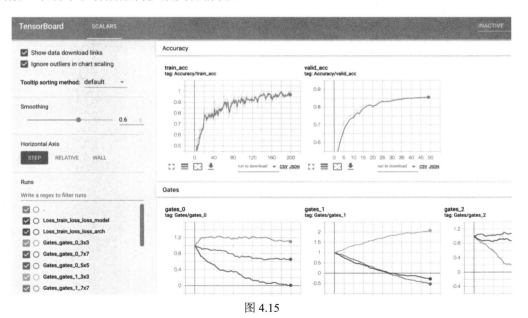

图 4.15

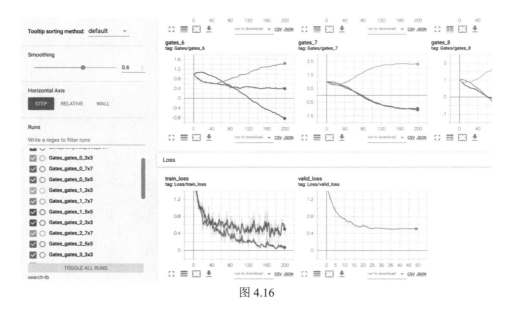

图 4.16

接下来便是利用训练好的超网络剪枝出搜索结构。这里我们取每一个 BasicBlockSearch 中 gates 最大值对应的 branch。以下实现包含在 prune_model.py 文件中。

```python
from model_search import ResNetSearch
import torch
import argparse
import os
import misc

parser = argparse.ArgumentParser()
parser.add_argument('--depth', default=20, type=int)

args = parser.parse_args()
args.num_classes = 10
args.model_weights_path = os.path.join(    # 待加载的训练好的超网络权重
    'logs/search-resnet%d' % args.depth,
    'model_search.pth'
)

model = ResNetSearch(args.depth, args.num_classes)
model.load_state_dict(torch.load(args.model_weights_path))
kernel_size = [3, 5, 7]
padding = [1, 2, 3]

config = []
for m in model.modules():                    # 搜索出对应 branch
    if m.__class__.__name__ == 'BasicBlockSearch':
        select_idx = m.gates.argmax().item()
        mid_channel = m.mid_channels[select_idx]
```

```
        config.append({
            'kernel_size': kernel_size[select_idx],
            'padding': padding[select_idx],
            'mid_channel': mid_channel
        })
print('Searched ResNet%d: ' % args.depth)
for c in config:
    print(c)

misc.dump_pickle(config, os.path.join(
    'logs/search-resnet%d' % args.depth,
    'model_config.pkl'
))
```

这里我们直接将搜索到的网络配置信息输出到终端。

```
$ python prune_model.py --depth 20
Searched ResNet20:
{'padding': 1, 'mid_channel': 16, 'kernel_size': 3}
{'padding': 2, 'mid_channel': 16, 'kernel_size': 5}
{'padding': 1, 'mid_channel': 16, 'kernel_size': 3}
{'padding': 1, 'mid_channel': 32, 'kernel_size': 3}
{'padding': 1, 'mid_channel': 32, 'kernel_size': 3}
{'padding': 1, 'mid_channel': 32, 'kernel_size': 3}
{'padding': 2, 'mid_channel': 64, 'kernel_size': 5}
{'padding': 1, 'mid_channel': 64, 'kernel_size': 3}
{'padding': 1, 'mid_channel': 64, 'kernel_size': 3}
Dumping pickle object to logs/search-resnet20/model_config.pkl
```

可以看到在本次实验中搜索出的网络结构中各层的配置。其中主要还是 3×3 卷积，但是在第 2 个和第 7 个 BasicBlock 选择了 5×5 卷积。

接下来便是利用搜索到的网络结构从头开始训练网络。这里网络结构搭建不再过多叙述，代码实现与 model_search.py 类似，只是不再使用多路分支加权得到输出。在训练模型时，这里使用 SGD 优化器，优化总周期为 160，使用 MultiStepLR 学习率调整器。数据预处理相比于之前稍有不同。我们使用了 Cutout[43]这个操作，使得输入图片某个随机区域被裁切，其效果如图 4.17 所示（取自原论文）。

图 4.17

具体实现也很简单，代码如下：

```
class Cutout(object):
    def __init__(self, length):          # 设置裁切区域长宽
        self.length = length

    def __call__(self, img):
        h, w = img.size(1), img.size(2)
        y = torch.randint(h, (1,))
        x = torch.randint(w, (1,))
                                         # 不超过图片边界
        y1 = torch.clamp(y - self.length // 2, 0, h)
        y2 = torch.clamp(y + self.length // 2, 0, h)
        x1 = torch.clamp(x - self.length // 2, 0, w)
        x2 = torch.clamp(x + self.length // 2, 0, w)

        img[:, y1: y2, x1: x2] = 0.
        return img
```

这样在按照以上搜索出网络结构和设置的超参数条件下，从头开始训练，最终可以在 CIFAR10 数据集上达到 92.68% 的测试准确率。而作为对比，原 ResNet20 模型达到 92.56% 测试准确率，而不使用 Cutout 数据预处理流程训练 ResNet20 模型则只有 91.88% 的准确率。因此可以看出我们搜索得到的网络结构要优于普通设计的结构，同时也验证了 Cutout 数据预处理的有效性。

## 4.5 PyTorch 应用实战二：在 ImageNet 数据集进行弱监督物体定位

在这一节中，我们将介绍在 ImageNet 数据集上进行弱监督物体定位任务。该任务使用到 GradCAM[44] 这种解释显图（Explanatory Saliency Map）方法，可以用于对输入图片进行分析，产生对于模型预测结果贡献重要性的显著图。这种显著图可以和图片中物体位置有较好的对应，因此可以用来做弱监督物体定位（Weakly Supervised Object Localization，WSOL）任务。最后我们将在 ImageNet 验证集上的 50 000 张图片进行实验。

### 4.5.1 GradCAM 解释显著图方法介绍

首先我们介绍产生解释显著图的方法 GradCAM。该方法继承于类别响应图法[45]（Class Activation Mapping，CAM）。该方法主要利用了深度学习网络最后分类层权重与类别语义对应的概念，在特征提取的最后输出特征上，寻找某种类别物体响应最高的区域。

CAM 方法如图 4.18 所示（取自原论文）。其假定网络经过一系列卷积运算后，得到

特征图（Feature Map）。然后经过全局平均池化（Global Average Pooling，GAP）之后，再与分类层权重相乘得到对于每类预测输出。而 CAM 的产生则是利用了预测某一类别的权重$\{w_1, w_2, \cdots, w_n\}$对于特征图按照每一个通道进行加权求和，得到二维的显著图。

但是以上方法要求网络中需要存在 GAP 环节，并且要求 GAP 后直接连接分类权重。这对于如 AlexNet 和 VGG16 等模型来说，其分类部分包含多个全连接层，并不能直接应用。GradCAM 则将 CAM 方法进行推广，其主要改变在于用来产生类别响应图的加权权重，变成了某一类别输出对于特征图的反传梯度向量。这个反传梯度与特征图尺寸相同，而为了按通道加权，则需要按通道求平均值，再逐通道加权求和。由于高层特征图往往经过了空间分辨率上的降采样操作，因此最终显示的时候，需要将其上采样为和输入图片相同大小尺寸。

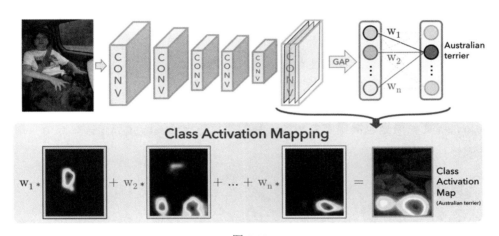

图 4.18

### 4.5.2 弱监督物体定位任务

以上介绍了产生显著图的方法。从图 4.18 中可以看出显著图响应强烈的地方，对应于预测物体类别的区域。因此虽然在训练网络过程中我们并没有提供其关于物体位置的信息，但是网络本身学习出了寻找图片中物体区域的特性，因此实现了弱监督信号（只提供了物体类别信息）下的物体定位任务。

而如何利用显著图产生最终的物体定位框（Bounding Box），这里我们采取一种最简单的策略。首先记图片产生的显著图为 $S$，利用阈值 $v = \alpha\mu$ 对显著图进行二值化，其中 $\mu$ 是显著图强度的均值。这样当高于阈值的区域置为 1，小于阈值的区域置为 0 时，则可以得到一个二值化掩码。然后选取包含掩码为 1 区域的最紧长方形框作为物体定位框，则完成了弱监督物体定位流程。该过程如图 4.19 所示。每一行从左到右显示了定位流程。上下两行分别代表了利用 GradCAM 对大象和斑马两类物体的定位情况。

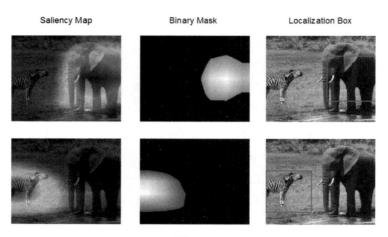

图 4.19

从图 4.19 中可以直观地看出所产生的显著图和定位框与物体真实位置较为准确。但是如何能够定量化描述这种准确性？这里引入在物体定位一种常用衡量标准，即如果所预测的定位框，与图片中真实物体定位框的交并比（Intersection Over Union，IOU）大于 0.5 时，则认为定位准确。我们所进行实验的数据集 ImageNet 图片由于每张图片中只有一类物体，因此如果定位准确则判别为这种图片预测准确，因此可以计算准确率来衡量定位准确性。

### 4.5.3 具体实现

接下来介绍整体任务的具体实现。首先是数据集准备，需要下载 ImageNet 验证集 50 000 张图片和其对应的每张图片物体真实位置框标注等信息。这里推荐使用以下链接：① https://onedrive.hyper.ai/down/ImageNet/data/ImageNet2012/ILSVRC2012_img_val.tar ；② https://raw.githubusercontent.com/clovaai/wsolevaluation/master/metadata/ILSVRC/test/class_labels.txt ；③ https://raw.githubusercontent.com/clovaai/wsolevaluation/master/metadata/ILSVRC/test/image_sizes.txt ；④ https://raw.githubusercontent.com/clovaai/wsolevaluation/master/metadata/ILSVRC/test/localization.txt。下载好图片后解压到 val 文件夹，然后利用 prepare_data.py 提取信息，代码为：

```
import misc
from collections import OrderedDict

metadata = OrderedDict()                          # 存储图片所有信息

with open('class_labels.txt', 'r') as f:          # 读取图片类别信息
    lines = f.readlines()

for line in lines:
    filename, label = line.strip('\n').split(',')
```

```python
        metadata[filename] = {'label': int(label)}

with open('image_sizes.txt', 'r') as f:               # 读取图片尺寸信息
    lines = f.readlines()

for line in lines:
    filename, width, height = line.strip('\n').split(',')
    metadata[filename]['image_size'] = [int(height), int(width)]

with open('localization.txt', 'r') as f:              # 读取图片定位框信息
    lines = f.readlines()

for line in lines:                                    # 可能包含多个定位框
    filename, x1, y1, x2, y2 = line.strip('\n').split(',')
    if metadata[filename].get('gt_bboxes') is None:
        metadata[filename]['gt_bboxes'] = [
            [int(x1), int(y1), int(x2), int(y2)]
        ]
    else:
        metadata[filename]['gt_bboxes'].append(
            [int(x1), int(y1), int(x2), int(y2)]
        )

misc.dump_pickle(metadata, 'imagenet_val_gt.pkl')     # 存储信息
```

这里的 misc.py 与 4.4 节中一样。运行以上程序，则会得到 imagenet_val_gt.pkl 文件，存储所需信息。然后便可以自定义数据集加载以上信息，代码在 dataset.py 中。

```python
from torch.utils import data
import misc
import torch


class ImageNet(data.Dataset):
    def __init__(self, metadata, transform):
        self.metadata = metadata                      # 图片所有信息
        self.transform = transform                    # 图片转换操作

    def __getitem__(self, item):
        index = '000%05d' % (item + 1)                # 图片名称编号
        name = 'val/ILSVRC2012_val_%s.JPEG' % index
        info = self.metadata[name]
        img = misc.pil_loader(name)                   # 加载图片
        img = self.transform(img)                     # 转换图片
        label = info['label']
        size = info['image_size']
        bbox = info['gt_bboxes']
        return img, label, size, bbox
```

```python
    def __len__(self):                              # 共50 000张图片
        return 50000


def custom_collate_fn(batch):                       # 由于自定义数据集返回格式不同
    imgs = []                                       # 需要自定义collate_fn
    labels = []
    sizes = []
    bboxes = []
    for img, label, size, bbox in batch:
        imgs.append(img)
        labels.append(label)
        sizes.append(size)
        bboxes.append(bbox)

    imgs = torch.stack(imgs)                        # 图片扩展为NCHW维度向量
    labels = torch.Tensor(labels).long()            # 标签为长整数型

    return imgs, labels, sizes, bboxes              # sizes和bboxes以列表返回
```

接下来是核心部分，实现 GradCAM 方法。我们需要获取给定模型某一层在前向计算过程中的输出特征以及反向回传时的梯度。这里使用 PyTorch 中提供的 register_hook 机制。所谓 hook 函数，指的是附加在模型某一层的一种记录函数，它可以记录下该层输入输出等信息，并提取出来供外部使用。这样的机制使得我们无需改动原模型结构即可以提取中间层信息。以下代码展示了这一用法，包含在 gradcam.py 中。

```python
import torch

def get_layer(model, key_list):                     # 根据名称获取模型中某一层
    a = model
    for key in key_list:
        a = a._modules[key]
    return a


class GradCAM(object):
    def __init__(self, model, target_layer):
        self.model = model
        self.target_layer = get_layer(
            model, target_layer
        )
        self.activations = []                       # 用来存储该层输出响应
        self.gradients = []                         # 用来存储该层反传梯度
        self.register_hooks()                       # 对某一层添加hook函数

    def register_hooks(self):
        def forward_hook(module, input, output):    # 前向hook函数，记录输出
            self.activations.append(output.data.clone())
```

```python
                                        # 反向 hook 函数，记录梯度
        def backward_hook(module, grad_input, grad_output):
            self.gradients.append(grad_output[0].data.clone())

        self.target_layer.register_forward_hook(forward_hook)
        self.target_layer.register_backward_hook(backward_hook)

    def reset(self):                    # 在每次重新产生 CAM 时，清空已有结果
        self.activations.clear()
        self.gradients.clear()

    def generate_saliency_map(self, input, label):
        self.reset()

        output = self.model(input)
        grad_output = output.data.clone() # 对输出给以 onehot 形式反传输入
        grad_output.zero_()
        grad_output.scatter_(1, label.unsqueeze(0).t(), 1.0)
        output.backward(grad_output)

        act = self.activations[0]        # 获取特征输出
        grad = self.gradients[0]         # 获取梯度

        weight = grad.mean(dim=(2, 3), keepdim=True)     # 按通道平均
        cam = (weight * act).sum(dim=1, keepdim=True)    # 计算 CAM
        cam = torch.clamp(cam, min=0)    # 截止为 0

        return cam
```

图 4.19 中的显著图结果即由以上代码生成，验证了准确性。最后是实现弱监督定位过程。这里先实现如何从显著图获取定位框的函数，以下代码包含在 helper.py 中。

```python
import numpy as np

def getbb_from_heatmap(hmp, alpha):          # 从显著图获取定位框
    heatmap = np.copy(hmp)

    threshold = alpha * heatmap.mean()
    heatmap[heatmap < threshold] = 0         # 产生二值化掩码

    if (heatmap == 0).all():                 # 如果均小于阈值，则默认全图为框
        return [[1, 1, heatmap.shape[0], heatmap.shape[1]]]

    x = np.where(heatmap.sum(0) > 0)[0] + 1  # 获取横纵坐标
    y = np.where(heatmap.sum(1) > 0)[0] + 1
    return [[x[0], y[0], x[-1], y[-1]]]
```

```python
def ious(pred, gt):                                    # 根据预测框和真实框计算IOU
    numObj = len(gt)
    gt = np.tile(gt,[len(pred),1])
    pred = np.repeat(pred, numObj, axis=0)
    bi = np.minimum(pred[:,2:],gt[:,2:]) - \
        np.maximum(pred[:,:2], gt[:,:2]) + 1
    area_bi = np.prod(bi.clip(0), axis=1)              # 计算交叉部分
    gt_area = (gt[:,2]-gt[:,0]+1) * (gt[:,3]-gt[:,1]+1)
    pred_area = (pred[:,2]-pred[:,0]+1) * (pred[:,3]-pred[:,1]+1)
    area_bu = gt_area + pred_area - area_bi            # 计算联合部分
    return area_bi / area_bu
```

最后是整体定位流程。在对显著图进行二值化时，需要设置阈值参数 $\alpha$，这里我们采取在一定区间 [0:alpha_step:max_alpha] 内遍历的方法，观察不同 $\alpha$ 对于定位性能的影响。同时我们还运行模型在 AlexNet、VGG16 和 ResNet50 模型上实验，观察不同网络结构对于定位性能的影响。由于图片数量众多且模型运算复杂，这里依然推荐使用 GPU 计算。代码包含在 localization.py 中。

```python
from dataset import ImageNet, custom_collate_fn
from gradcam import GradCAM
from torchvision import models, transforms
from torch.utils import data
from tqdm import tqdm                                  # 在终端显示进度条
import torch.nn.functional as F
import numpy as np
import misc
import helper
import torch
import argparse
import os

print = misc.logger.info

parser = argparse.ArgumentParser()                     # 超参数设置
parser.add_argument('--gpu', default='0', type=str)
parser.add_argument('--arch', default='alexnet', type=str)
parser.add_argument('--batch_size', default=100, type=int)
parser.add_argument('--max_alpha', default=5.2, type=float)
parser.add_argument('--alpha_step', default=0.2, type=float)

args = parser.parse_args()
args.device = 'cuda'
torch.backends.cudnn.benchmark = True
os.environ['CUDA_VISIBLE_DEVICES'] = args.gpu

if args.arch == 'alexnet':                             # 提取最后MaxPool输出
    args.target_layer = ['features', '12']
```

```python
elif args.arch == 'vgg16':
    args.target_layer = ['features', '30']   # 提取最后 MaxPool 输出

elif args.arch == 'resnet50':                # 提取最后 Bottleneck 模块输出
    args.target_layer = ['layer4', '2']

else:
    raise NotImplementedError

args.logdir = 'wsol-%s' % args.arch

misc.prepare_logging(args)

tfm = transforms.Compose([                   # 图片处理变换
    transforms.Resize((224, 224)),
    transforms.ToTensor(),
    transforms.Normalize(
        mean=[0.485, 0.456, 0.406],
        std=[0.229, 0.224, 0.225]
    )
])

print('==> Preparing data..')

metadata = misc.load_pickle(                 # 加载定位框等信息
    'imagenet_val_gt.pkl'
)
data_loader = data.DataLoader(               # 加载数据流
    ImageNet(metadata, tfm),
    batch_size=args.batch_size,
    collate_fn=custom_collate_fn,
    num_workers=4, pin_memory=True
)

print('==> Initializing model...')

model = models.__dict__[args.arch](          # 加载预训练模型
    pretrained=True
)
model.eval()
model.to(args.device)
explainer = GradCAM(model, args.target_layer) # 定义GradCAM解释显著图生成器

num_alphas = len(np.arange(0, args.max_alpha, args.alpha_step))
correct_loc = torch.zeros(num_alphas)        # 记录每个 alpha 对应定位准确率

for data, label, sizes, gt_bboxes in tqdm(data_loader):
    data, label = data.to(args.device), label.to(args.device)
```

```python
    cam = explainer.generate_saliency_map(    # 产生原始 CAM
        data, label
    )

    for i in range(len(data)):
        saliency_map = F.interpolate(          # 每个显著图放大到对应图片大小
            cam[i].unsqueeze(0), sizes[i], mode='bilinear'
        )
        saliency_map = saliency_map.squeeze().cpu().numpy()
        alphas = np.arange(0, args.max_alpha, args.alpha_step)
        for k, alpha in enumerate(alphas):
            pred_bbox = helper.getbb_from_heatmap(     # 获取定位框
                saliency_map, alpha
            )
            ious = helper.ious(pred_bbox, gt_bboxes[i])  # 计算 IOU
            if max(ious) > 0.5:                          # 如果大于 0.5
                correct_loc[k] += 1                      # 则判断定位正确

correct_loc /= len(data_loader.dataset)                  # 计算定位准确率
alphas = np.arange(0, args.max_alpha, args.alpha_step)
for k, alpha in enumerate(alphas):                       # 输出信息
    print('loc acc = %.4f @ alpha = %.2f' % (
        correct_loc[k].item(), alpha
    ))
misc.dump_pickle(correct_loc, os.path.join(args.logdir, 'loc_acc.pkl'))
```

如果以上实现正确，运行代码则可以得到如下的输出结果。

```
$ python localization.py --gpu 0 --arch resnet50
[15:51:15.481] =================FLAGS=================
[15:51:15.482] alpha_step: 0.2
[15:51:15.482] batch_size: 100
[15:51:15.482] logdir: ./logs/wsol-resnet50
[15:51:15.482] gpu: 0
[15:51:15.482] max_alpha: 5.2
[15:51:15.482] device: cuda
[15:51:15.482] arch: resnet50
[15:51:15.482] target_layer: ['layer4', '2']
[15:51:15.482] =======================================
[15:51:15.482] ==> Preparing data..
Loading pickle object from imagenet_val_gt.pkl
=> Done (0.1962 s)
[15:51:15.679] ==> Initializing model...
100%|████████████████████████████████████████
| 500/500 [20:02<00:00, 2.40s/it]
[16:11:23.518] loc acc = 0.4075 @ alpha = 0.00
[16:11:23.520] loc acc = 0.4291 @ alpha = 0.20
[16:11:23.520] loc acc = 0.4704 @ alpha = 0.40
```

```
[16:11:23.520] loc acc = 0.5169 @ alpha = 0.60
[16:11:23.520] loc acc = 0.5576 @ alpha = 0.80
[16:11:23.520] loc acc = 0.5843 @ alpha = 1.00
[16:11:23.521] loc acc = 0.5595 @ alpha = 1.20
[16:11:23.521] loc acc = 0.4501 @ alpha = 1.40
[16:11:23.521] loc acc = 0.3416 @ alpha = 1.60
[16:11:23.521] loc acc = 0.2726 @ alpha = 1.80
...
...
```

图 4.20 总结了 ImageNet 数据集上不同模型在不同 $\alpha$ 情况下的定位准确率曲线，并且标出了不同模型最大定位准确率水平。可以看出 ResNet50 最高可以到达 58.43%，VGG16 次之，AlexNet 最低。这也和三种模型在 ImageNet 上图片分类准确率水平高低一致。说明了更好的特征提取模型也蕴含着更准确的物体定位。而探究二者之间的因果关系，以及利用更复杂的解释显著图方法和定位框提取方法等内容，留待各位读者自行探索。

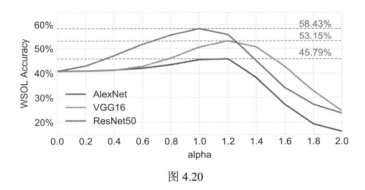

图 4.20

# 第 5 章 PyTorch 高级技巧与实战应用

经过了前面几章的学习，我们已经掌握了利用 PyTorch 构建较为复杂应用的整体流程。但是所应用的场景主要还局限于单机单卡情况下，处理中小规模的数据。但是在一些真实应用场景及最前沿的科学研究中，利用大规模数据及 GPU 设备进行并行计算是常出现的情况。同时之前的应用更多是基于 PyTorch 提供的 Python 接口展开，如何从底层 C++ 和 CUDA 实现自定义算子的扩展，如何将基于 Python 环境训练好的模型导出，并在其他语言如 C++ 中加载，是工程实践中面临的真实问题。在这一节中我们将对以上内容作全面介绍。

## 5.1 PyTorch 并行计算

在本节中，我们将介绍 PyTorch 并行计算应用范例，其中涉及到如何分布式加载大规模数据，如何调度使用高效的 GPU 并行计算过程，如何进一步加速模型计算，减少显存使用量。在本章最后的应用实战环节，我们将这些内容综合在一起，全面应用于在 ImageNet 数据集上完整训练 MobileNetV2 的过程。

### 5.1.1 大规模数据集加载

在面对大规模数据情况下，全部将数据加载至内存后进行迭代往往既不可实现，也不可忍受，因为需要消耗大量的内存与时间。在这种情况下只能是先获取每个数据样本的索引路径，在真正迭代时现场从存储空间加载至内存。但是由于机器学习模型训练过程中往往要求数据随机化，这种频繁的从存储空间随机读取比较耗时。面对这种情况有多种技术实践，比如使用 LMDB 格式的数据库存储文件，或者如 TensorFlow 中推荐的 tfrecord 格式将数据分割为多个片段。但是这些存储格式往往限制了数据随机性，因为其无法进行全局单例样本级别的打乱重排操作。在作者经过多种实践探索后，最简单省心的解决方案就是使用固态硬盘(SSD)，其随机读写等操作速度远远快于机械硬盘(HDD)。这种从硬件层面本质级别的加速远胜于软件调度带来的提升。

在此基础之上，还需要构建适配于分布式训练的数据流。PyTorch 中的分布式计算框架是多进程操作，为每个 GPU 计算单元分配一个单独进程进行数据读取。因此相应的数据采样过程需要改变。这里使用 torch.utils.data.distributed.DistributedSampler。之前在 4.1.2 节介绍过，Sampler 负责采样索引，传递至 Dataset 类中，再由 DataLoader 构成数据流。DistributedSampler 则是处理分布式多进程中的索引采样环节。这里我们可以研读其源码，体会实现过程。

```python
import math
import torch
from . import Sampler
import torch.distributed as dist


class DistributedSampler(Sampler):

    def __init__(self, dataset, num_replicas=None, rank=None, shuffle=True):
        if num_replicas is None:
            if not dist.is_available():
                raise RuntimeError(
                    "Requires distributed package to be available"
                )
            num_replicas = dist.get_world_size()   # 获取总进程数
        if rank is None:
            if not dist.is_available():
                raise RuntimeError(
                    "Requires distributed package to be available"
                )
            rank = dist.get_rank()                 # 获取本进程编号
        self.dataset = dataset
        self.num_replicas = num_replicas
        self.rank = rank
        self.epoch = 0
        self.num_samples = int(math.ceil(          # 计算每个进程分配样本数
            len(self.dataset) * 1.0 / self.num_replicas
        ))
        self.total_size = self.num_samples * self.num_replicas  # 总样本量
        self.shuffle = shuffle

    def __iter__(self):
        # deterministically shuffle based on epoch
        g = torch.Generator()
        g.manual_seed(self.epoch)                  # 根据周期设置随机种子
        if self.shuffle:                           # 重排索引
            indices = torch.randperm(
                len(self.dataset), generator=g
            ).tolist()
        else:
            indices = list(range(len(self.dataset)))


        # add extra samples to make it evenly divisible    # 索引数达到总样本量
        indices += indices[:(self.total_size - len(indices))]
        assert len(indices) == self.total_size
```

```
        # subsample                                    # 每个进程采样索引
        indices = indices[self.rank:self.total_size:self.num_replicas]
        assert len(indices) == self.num_samples

        return iter(indices)

    def __len__(self):
        return self.num_samples

    def set_epoch(self, epoch):                         # 设置周期数
        self.epoch = epoch
```

从上面的实现中我们可以看出，利用 torch.distributed 模块，在获取每个进程编号和总进程数目之后，将索引分割为互不重叠的部分即可。这里引入了 torch.distributed 模块，提供了处理多进程通讯的基本原语（primitive）。除了以上的获取进程数和进程编号操作，还支持点对点操作 send/recv、集合操作 scatter/gather/reduce 等等。其背后实现利用的三种后端（backend）分别为 gloo、mpi 和 nccl。一般来说对于分布式 GPU 进程间通讯使用 nccl 后端，对于分布式 CPU 进程间通讯使用 gloo 后端。这些原语操作也是之后支持 GPU 并行计算的基础。

在设计大规模数据流时，除了以上的分布式采样器之外，还需要关注预处理部分。一般对于图像的预处理流程计算较为简单时，可以完全放置于 CPU 上进行处理。而如果是针对于如视频解码等计算耗时较多的操作，此处推荐是 NVIDIA 开发的 DALI[①] 库，利用 GPU 进行数据预处理。

在本书 4.1 节中介绍 DataLoader 构成时，提到了 collate_fn 概念，用来将多个样本整合为一个 batch。PyTorch 中默认的 collate_fn 包含了一些冗余的分类和检查。如果在已知数据类型情况下，可以去除这些步骤，编写自定义的 fast_collate_fn 如下[②]：

```
def fast_collate_fn(batch):
    imgs = [img[0] for img in batch]
    targets = torch.tensor([
        target[1] for target in batch
    ], dtype=torch.int64)
    w = imgs[0].size[0]
    h = imgs[0].size[1]
    tensor = torch.zeros((len(imgs), 3, h, w), dtype=torch.uint8)
    for i, img in enumerate(imgs):
        nump_array = np.asarray(img, dtype=np.uint8)
        if(nump_array.ndim < 3):
            nump_array = np.expand_dims(nump_array, axis=-1)
        nump_array = np.rollaxis(nump_array, 2)
```

---

① https://github.com/NVIDIA/DALI

② 节选自 https://github.com/NVIDIA/apex/blob/master/examples/imagenet/main_amp.py

```
            tensor[i] += torch.from_numpy(nump_array)

    return tensor, target
```

这里传入的 img 是直接用 PIL.Image 读取后的输入，而转换成 torch.Tensor 的步骤则是整体进行，减少了每个进程对单独样本的转换。

最后在进行迭代产生数据样本时，PyTorch 的 DataLoader 已经实现了预读取（prefetch）过程，即在模型运算同时加载，处理数据并迁移至 GPU 上，从而弥补掉这部分耗时。而在 Apex 给出的示例代码中①，此处可以通过额外建立 torch.cuda.Stream 方式，进一步将这部分操作与模型计算进行并行，从而减少时间。

```
class DataPrefetcher():
    def __init__(self, loader):
        self.loader = iter(loader)
        self.stream = torch.cuda.Stream()
        self.mean = torch.tensor(                       # 归一化用到的均值和标准差
            [0.485 * 255, 0.456 * 255, 0.406 * 255]
        ).cuda().view(1,3,1,1)
        self.std = torch.tensor(
            [0.229 * 255, 0.224 * 255, 0.225 * 255]
        ).cuda().view(1,3,1,1)
        self.preload()

    def preload(self):                                  # 预加载部分
        try:
            self.next_input, self.next_target = next(self.loader)
        except StopIteration:
            self.next_input = None
            self.next_target = None
            return
        with torch.cuda.stream(self.stream):            # 利用额外 Stream
            self.next_input = self.next_input.cuda(async=True)
            self.next_target = self.next_target.cuda(async=True)
            self.next_input = self.next_input.float()
            self.next_input = self.next_input.sub_(self.mean).div_(self.std)

    def next(self):
        torch.cuda.current_stream().wait_stream(self.stream)
        input = self.next_input
        target = self.next_target
        self.preload()
        return input, target
```

---

① https://github.com/NVIDIA/apex/blob/master/examples/imagenet/main_amp.py

有关于该 DataPrefetcher 加速情况的探讨可以参考这个 issue 链接[①]，以及 PyTorch 官方论坛[②]。

### 5.1.2　模型的高效并行计算

上一节主要介绍了分布式数据加载方式。这一节将介绍模型的高效并行计算。在深度模型训练过程当中，存在这两种模型并行计算方式：一种是数据并行（DataParallel），即每个计算单元复制相同的模型，数据输入均分到各个单元上，计算结束后再统一综合出结果；另一种是模型并行（ModelParallel），指的是将模型各个部分分配到不同计算单元上，这主要处理一些模型内部存在并行计算的结构，或是模型本身很大无法在单个计算单元上容纳的场景。由于数据并行计算逻辑简单，因此在 PyTorch 中主要支持数据并行的计算方式。其中提供了两个接口。第一种是 nn.DataParallel，这是早期的数据并行方式，但这种方式并未做到多进程，而是在将多 GPU 的运行结果合并至 master GPU 上，进行累加，再广播到各个计算单元上。这会造成 master GPU 成为整个系统的计算瓶颈。另一种方式是 nn.DistributedDataParallel，是当前 PyTorch 推荐的数据并行计算方式，因为它对每个 GPU 都分配了单独的进程，从而避免了因 Python GIL 限制产生的计算开销。

使用 nn.DistributedDataParallel 很简单，大致流程如下：

```python
import torch
import torch.nn.DistributedDataParallel as DDP

# args是命令行传入参数
args.gpu = args.local_rank % torch.cuda.device_count()  # 进程对应的GPU卡号
torch.cuda.set_device(args.gpu)
torch.distributed.init_process_group(backend='nccl',    # 初始化进程组
                                     init_method='env://')
args.world_size = torch.distributed.get_world_size()    # 获取总体进程数
...
model = ModelInitialization()
model = model.cuda()
model = DDP(model, device_ids=[args.gpu])               # 建立分布式模型
```

之后的使用如单进程的用法一样，例如前向计算、反向求导、更新参数等等。这其中主要利用内部实现的参数梯度规约求和的操作。在此过程中参数从未被广播，只是将各个进程的梯度进行了求和通讯，然后再广播到各个进程，利用每个进程的优化器更新参数。而模型中 Buffer 的内容（如 BatchNorm 层中统计的均值方差），则是由 0 号进程广播到各个节点。这样的操作可以使得运算部分依然是分布式进行，主要的开销在于求和和广播部分。需要注意的是，在获取模型输出结果时，应当对多个进程的结果求平均，

---

[①]　https://github.com/NVIDIA/apex/issues/304#

[②]　https://discuss.pytorch.org/t/how-to-prefetch-data-when-processing-with-gpu/548/19

可使用如下函数：

```
import torch.distributed as dist
def reduce_tensor(tensor):
    rt = tensor.clone()
    dist.all_reduce(rt, op=dist.reduce_op.SUM)   # 求和
    rt /= args.world_size                         # 求平均
    return rt
```

这里使用到了 torch.distributed 中的规约操作。更多分布式操作我们以实现 SyncBatchNorm 过程作为示例。该函数实现了在分布式计算模型中，各个进程模型中的 BatchNorm 层同步统计均值方差的操作。这种同步后的 BatchNorm 层在物体检测任务中使用，因为往往此时各个进程的 batch size 较小，如果只是依赖于主进程的统计值，则会产生严重的偏差，而同步之后相当于在更大的 batch size 上进行统计，使得训练更稳定更容易收敛。PyTorch 官方实现的 SyncBatchNorm 函数如下[①]：

```
import torch
from torch.autograd.function import Function

class SyncBatchNorm(Function):

    @staticmethod
    def forward(self,
                input,
                weight,
                bias,
                running_mean,
                running_var,
                eps,
                momentum,
                process_group,
                world_size):
        input = input.contiguous()

        count = torch.empty(1, dtype=running_mean.dtype,
            device=input.device).fill_(input.numel() // input.size(1))

        # 计算输入的 mean/invstd
        mean, invstd = torch.batch_norm_stats(input, eps)

        count_all = torch.empty(
            world_size, 1, dtype=count.dtype,
            device=count.device
        )
        mean_all = torch.empty(
```

---

① https://github.com/pytorch/pytorch/blob/master/torch/nn/modules/_functions.py

```python
        world_size, mean.size(0),
        dtype=mean.dtype, device=mean.device
)
invstd_all = torch.empty(
        world_size, invstd.size(0),
        dtype=invstd.dtype, device=invstd.device
)

count_l = list(count_all.unbind(0))       # 按照维度 0 解耦成为列表
mean_l = list(mean_all.unbind(0))
invstd_l = list(invstd_all.unbind(0))

# 使用 all_gather 操作
# 一次性计算 count/mean/var
count_all_reduce = torch.distributed.all_gather(
        count_l, count, process_group, async_op=True
)
mean_all_reduce = torch.distributed.all_gather(
        mean_l, mean, process_group, async_op=True
)
invstd_all_reduce = torch.distributed.all_gather(
        invstd_l, invstd, process_group, async_op=True
)

# 等待异步通信结束
count_all_reduce.wait()
mean_all_reduce.wait()
invstd_all_reduce.wait()

size = count_all.view(-1).long().sum()

# 计算全局 mean 和 invstd
mean, invstd = torch.batch_norm_gather_stats_with_counts(
    input,
    mean_all,
    invstd_all,
    running_mean,
    running_var,
    momentum,
    eps,
    count_all.view(-1)
)

self.save_for_backward(input, weight, mean, invstd, count_all)
self.process_group = process_group

# 进行逐点归一化
out = torch.batch_norm_elemt(input, weight, bias, mean, invstd, eps)
```

```
    return out
```

完整的分布式教程和文档可参考官网链接①②。

## 5.1.3 加速模型计算和减少显存使用

以上介绍的分布式计算已经可以达到一般任务的加速目的。那么如何进一步提高计算效率或许要从另外角度去考察。之前的计算默认都是基于 32 位浮点值进行的，而实际上对于机器学习模型来说，由于其训练过程引入的随机性，以及其各自我学习纠正的能力，并不一定完全需要精确运算。而如果能够将 32 位浮点值降低为半精度即 16 位进行运算，则可以大大减少计算开销，同时减小模型整体显存的占用。

事实上这个思路是可行的，此处我们介绍 NVIDIA 推出的整体混合精度（mixed precision）训练方案。所谓混合精度，即模型不能完全基于半精度值进行训练。因为实验中发现，由于 16 位表示的数值范围小于 32 位浮点值，因此原来在 32 位表示绝对值小于 1e-06 量级的梯度值在 16 位表示下统一成为 0，这样在本该多次迭代累积梯度的权重值，一直为零无法得到训练。

为此 NVIDIA 提出了混合精度训练流程，即保存一份权重的 32 位浮点精度表示副本，在每次前向反向计算时，转换为 16 位半精度值。这样实际运算开销还是被减少的。而为了解决过小梯度无法表示的问题，引入了 loss scale 的概念，即在反传前将 loss 扩大 S 倍，反传求得梯度后再缩小 S 倍（此处为 32 位精度除法运算）。然后将此梯度更新于 32 位浮点精度的权重副本，构成一次优化迭代循环。更多的讲解参考其开发文档③。

同时 NVIDIA 还推出了 apex 库，专门针对 PyTorch 框架实现了以上混合精度训练流程。其使用方法也非常简单，类似于如下代码：

```
model = ModelInitialization(...)
optimizer = optim.SomeOptimizer(model.parameters(), ...)

opt_level = 'O1'
loss_scale = 128.0
model, optimizer = amp.initialize(
    model, optimizer, opt_level=opt_level, loss_scale=loss_scale
)
...
loss = criterion(model(input), targets)
optimizer.zero_grad()

with amp.scale_loss(loss, optimizer) as scaled_loss:
    scaled_loss.backward()
```

---

① https://pytorch.org/docs/stable/nn.html#torch.nn.parallel.DistributedDataParallel
② https://pytorch.org/docs/stable/distributed.html
③ https://docs.nvidia.com/deeplearning/performance/mixed-precision-training/index.html

```
optimizer.step()
...
```

这里首先按照正常方式初始化模型和优化器，然后设定 opt_level（推荐使用 O1 级别）及 loss_scale，利用 apex.amp 包装模型和优化器。在此过程中，将对模型中浮点运算部分逐一替换成对应的半精度运算函数，而且同时在 optimizer 中维护全精度权重副本。在后续的计算过程中，使用 amp.scale_loss 上下文管理器，进行梯度缩放并更新权重。更多使用指南可参考 apex 库的 github 链接[①]，同时我们也将在本章末尾实战环节进行展示。

## 5.2 扩展 PyTorch

在这一节中，我们将介绍如何对现有 PyTorch 框架进行扩展。一方面随着各种最新研究进展成果出现，一些复杂灵活的运算子被开发出来。虽然其可以分解成为一系列 PyTorch 基本运算的组合，但是整体运行时复杂度很高，如何将其从底层上实现成为一个完整的基本算子是亟待解决的问题。另一方面，之前介绍的模型搭建计算更多关注于训练阶段，而在实际应用中如何部署 PyTorch 中的模型，如何在生产环境中与其他常使用的语言如 C++进行交互也是业界最关心的问题。在本节中我们将对以上方面的问题进行介绍。但应注意，限于篇幅，这部分内容我们只以使用者角度去理解，不做深入框架源码级的解读。更多探索内容读者可以自行在官方开源代码中学习。

### 5.2.1 利用 C++和 CUDA 实现自定义算子

在一些复杂应用中由于论文创新会提出一些复杂的算子。这些算子绝大部分还是可以依靠 PyTorch 中基本张量操作以及一些高级运算操作组合得到。但是这种组合往往包含了冗余数据拷贝和并行度较低的运算过程，导致在进行大规模数据训练时，造成如 GPU 计算效率低下，整体时间过长的结果。这时在排查出造成延时增加的主要环节是这些复杂算子导致之后，可以利用更加底层的接口和语言来整体实现算子功能。这里我们就介绍如何在 PyTorch 中利用 C++/CUDA 混合编程方式实现自定义算子。

我们以一个简单例子展示整体流程。这里实现一个称之为 negative concatenated ReLU 的激活函数，简称 NCReLU。其运算过程也很简单，对于输入 $x$，其输出为：

$$NCReLU(x) = concat(ReLU(x), -ReLU(-x))$$

其实现目的就是不仅在输出中将输入大于 0 的部分保留下来，也将小于 0 的部分一并保留，并且按照输出通道维度进行拼接。像类似于这种拼接激活函数，首次提出于论文[46]中，而在论文[47]中提出 NCReLU 对此进行改进，都是希望去补偿 ReLU 激活函数可能损失的信息。

---

[①] https://github.com/NVIDIA/apex

这里我们将这个激活函数实现为一个整体算子，编写在 ncrelu.cpp 文件中。然后利用 Python 中提供的 setuptools 模块完成事先编译流程，将写有算子的 C++文件，编译成为一个动态链接库（在 Linux 平台是一个.so 后缀文件），可以让 Python 调用其中实现的函数功能。为了编译 C++文件，首先需要编写如下 setup.py 文件：

```python
from setuptools import setup
from torch.utils import cpp_extension

setup(
    name='ncrelu_cpp',                          # 编译后的链接库名称
    ext_modules=[
        cpp_extension.CppExtension(
            'ncrelu_cpp', ['ncrelu.cpp']        # 待编译文件，及编译函数
        )
    ],
    cmdclass={                                  # 执行编译命令设置
        'build_ext': cpp_extension.BuildExtension
    }
)
```

这里 PyTorch 提供了一个封装 cpp_extension，方便设置编译过程中所需要的选项，以及所需包含的头文件位置的路径等等。接下来便是具体实现 ncrelu.cpp。

```cpp
#include <torch/extension.h>                    // 头文件引用部分

torch::Tensor ncrelu_forward(torch::Tensor input) {
    auto pos = input.clamp_min(0);              // 具体实现部分
    auto neg = input.clamp_max(0);
    return torch::cat({pos, neg}, 1);
}

PYBIND11_MODULE(TORCH_EXTENSION_NAME, m) {    // 绑定部分
  m.def("forward", &ncrelu_forward, "NCReLU forward");
}
```

以上代码包含了三个部分。首先是头文件引用部分。这里包含了 torch/extension.h 头文件，是编写 PyTorch 的 C++扩展时必须包含的一个文件。它基本上囊括了实现中所需要的所有依赖，包含了 ATen 库、pybind11 以及二者之间的交互。其中 ATen 是 PyTorch 底层张量运算库，负责实现具体张量操作运算；pybind11 是实现 C++代码到 Python 的绑定（binding），可以在 Python 里调用 C++函数。

第二部分是具体实现部分。函数返回类型和传递参数类型均是 torch::Tensor 类，这种对象不仅包含了数据，还附属了诸多运算操作。因此我们可以看到在下面的实现方式类似于 Python 中使用 PyTorch 张量运算操作一样，可以直接调用如截取操作 clamp 和拼接操作 torch::cat 等，非常简洁易读且方便。

第三部分就是绑定部分。只需要在 m.def 中传入参数，分别是绑定到 Python 的函数名称、需绑定的 C++函数引用，以及一个简短的函数说明字符串，用来添加到 Python 函数中的__doc__成员名称中。

将以上两个文件放在同一文件夹下，然后进行编译。这里可以运行 python setup.py install 命令，则是不仅编译出了动态链接文件，还将其拷贝至 Python 运行环境中的 /lib/python3.5/site-packages/中，可以全局导入使用。而我们这里只编译动态链接文件，并且只在该文件夹下使用，可以使用 python setup.py build_ext --inplace 命令。如果运行成功将会看到以下输出信息：

```
$ python setup.py build_ext --inplace
running build_ext
building 'ncrelu_cpp' extension
gcc -pthread -Wsign-compare -DNDEBUG -g -fwrapv -O3
    -Wall -Wstrict-prototypes -fPIC
    -I/PATH/TO/torch/include
    -I/PATH/TO/torch/include/torch/csrc/api/include
    -I/PATH/TO/torch/include/TH
    -I/PATH/TO/torch/include/THC
    -I/PATH/TO/include/python3.5m
    -c ncrelu.cpp
    -o build/temp.linux-x86_64-3.5/ncrelu.o -D
TORCH_API_INCLUDE_EXTENSION_H -DTORCH_EXTENSION_NAME=ncrelu_cpp
    -D_GLIBCXX_USE_CXX11_ABI=0 -std=c++11
creating build/lib.linux-x86_64-3.5
g++ -pthread -shared -L/PATH/TO/lib -Wl,-rpath=/PATH/TO/lib,
    --no-as-needed build/temp.linux-x86_64-3.5/ncrelu.o
    -L/PATH/TO/lib
    -lpython3.5m
    -o build/lib.linux-x86_64-3.5/ncrelu_cpp.cpython-35m-x86_64-linux-gnu.so
copying build/lib.linux-x86_64-3.5/ncrelu_cpp.cpython-35m-x86_64-linux-gnu.so ->
```

这里的/PATH/TO 是运行环境的链接引用路径，与个人设置有关。如果一切正常将会在文件夹下产生 ncrelu_cpp.cpython-35m-x86_64-linux-gnu.so 动态链接文件。我们可以启动 Python 检测是否可以导入其中的函数。

```
>> import torch
>> import ncrelu_cpp
>> a = torch.randn(10, 3)
>> a
tensor([[-1.0154,  0.7127, -0.9450],
        [ 0.2649, -0.1355, -0.1017],
        [-0.2768,  0.3547,  0.2203],
        [-1.4406, -2.1471, -0.3900],
        [ 1.1450,  0.2849, -0.9568],
        [-0.1243, -0.2891,  0.3181],
        [ 1.0269,  0.6391, -0.0657],
```

```
           [-0.8530,  0.3710, -1.2015],
           [ 1.4406, -1.3073, -0.6814],
           [-1.0122, -1.3921,  0.5631]]])
>> b = ncrelu_cpp.forward(a)
>> b
tensor([[ 0.0000,  0.7127,  0.0000, -1.0154,  0.0000, -0.9450],
        [ 0.2649,  0.0000,  0.0000,  0.0000, -0.1355, -0.1017],
        [ 0.0000,  0.3547,  0.2203, -0.2768,  0.0000,  0.0000],
        [ 0.0000,  0.0000,  0.0000, -1.4406, -2.1471, -0.3900],
        [ 1.1450,  0.2849,  0.0000,  0.0000,  0.0000, -0.9568],
        [ 0.0000,  0.0000,  0.3181, -0.1243, -0.2891,  0.0000],
        [ 1.0269,  0.6391,  0.0000,  0.0000,  0.0000, -0.0657],
        [ 0.0000,  0.3710,  0.0000, -0.8530,  0.0000, -1.2015],
        [ 1.4406,  0.0000,  0.0000,  0.0000, -1.3073, -0.6814],
        [ 0.0000,  0.0000,  0.5631, -1.0122, -1.3921,  0.0000]])
```

这里从结果可以看出，所实现的 ncrelu.cpp 中的 forward 函数正确地将输入中大于 0 和小于 0 的两部分拼接在了一起。而且由于我们是利用 PyTorch 中 ATen 张量库封装的高层操作，是一种与运行设备无关的代码抽象，因此上面所实现的函数可以直接应用于 GPU 上进行计算，只需要将输入迁移至 GPU 上即可。

```
>> a = a.cuda()
>> c = ncrelu_cpp.forward(a)
>> c
tensor([[ 0.0000,  0.7127,  0.0000, -1.0154,  0.0000, -0.9450],
        [ 0.2649,  0.0000,  0.0000,  0.0000, -0.1355, -0.1017],
        [ 0.0000,  0.3547,  0.2203, -0.2768,  0.0000,  0.0000],
        [ 0.0000,  0.0000,  0.0000, -1.4406, -2.1471, -0.3900],
        [ 1.1450,  0.2849,  0.0000,  0.0000,  0.0000, -0.9568],
        [ 0.0000,  0.0000,  0.3181, -0.1243, -0.2891,  0.0000],
        [ 1.0269,  0.6391,  0.0000,  0.0000,  0.0000, -0.0657],
        [ 0.0000,  0.3710,  0.0000, -0.8530,  0.0000, -1.2015],
        [ 1.4406,  0.0000,  0.0000,  0.0000, -1.3073, -0.6814],
        [ 0.0000,  0.0000,  0.5631, -1.0122, -1.3921,  0.0000]],
       device='cuda:0')
```

可以看出结果依然正确，并且所得结果张量的 device 属性为'cuda:0'。

以上所介绍的流程是先编译后加载的方式使用 C++实现扩展。而对于像本示例中这种较为简短的函数实现，PyTorch 中还支持另外一种即时（just-in-time，JIT）编译内联（inline）加载方式，即在一个 Python 文件中，将 C++代码整体作为字符串传递给 PyTorch 中负责内联编译的函数，然后在运行 Python 文件时，即时编译出动态链接文件，并导入其中函数进行后续运算。之前所实现的 NCReLU 可以用以下方式实现：

```
import torch
import torch.utils.cpp_extension

source = """                                          # C++源代码
```

```
#include <torch/extension.h>

torch::Tensor ncrelu_forward(torch::Tensor input) {
    auto pos = input.clamp_min(0);
    auto neg = input.clamp_max(0);
    return torch::cat({pos, neg}, 1);
}

PYBIND11_MODULE(TORCH_EXTENSION_NAME, m) {
  m.def("forward", &ncrelu_forward, "NCReLU forward");
}
"""
torch.utils.cpp_extension.load_inline(          # 内联加载
    name="ncrelu_cpp",
    cpp_sources=source,                         # 指定源码
    is_python_module=False,                     # 设置为非 python 模块
    verbose=True,                               # 输出编译信息
)

import ncrelu_cpp                               # 直接导入使用
a = torch.randn(10, 3)
print(a)
b = ncrelu_cpp.forward(a)
print(b)
```

在运行之前需要安装 ninja 库（pip install ninja），会在一个临时文件夹中编译出动态链接库。如果一切依赖正常，运行上面的代码则会输出以下信息：

```
Using /tmp/torch_extensions as PyTorch extensions root...
Emitting ninja build file /tmp/torch_extensions/ncrelu_cpp/build.ninja...
Building extension module ncrelu_cpp...
[1/2] c++ -MMD -MF main.o.d -DTORCH_EXTENSION_NAME=ncrelu_cpp
    -DTORCH_API_INCLUDE_EXTENSION_H -isystem /PATH/TO/torch/include
    -isystem /PATH/TO/torch/include/torch/csrc/api/include
    -isystem /PATH/TO/torch/include/TH
    -isystem /PATH/TO/torch/include/THC
    -isystem /PATH/TO/include/python3.5m
    -D_GLIBCXX_USE_CXX11_ABI=0 -fPIC -std=c++11
    -c /tmp/torch_extensions/ncrelu_cpp/main.cpp
    -o main.o
[2/2] c++ main.o -shared -o ncrelu_cpp.so
Loading extension module ncrelu_cpp...
tensor([[-0.0693,  0.9379,  1.5714],
        [ 1.8012,  2.8181, -0.5107],
        [ 1.4348, -1.7046,  2.1390],
        [ 0.7165, -0.0721, -0.2628],
        [ 0.5234,  0.0077,  0.6439],
        [ 0.7457, -1.5133,  1.6667],
        [-0.8872,  1.6385, -0.2752],
```

```
                [-0.6760, -0.1987, -0.1372],
                [ 1.2641,  0.4682,  0.8235],
                [-1.8230, -0.1598,  0.4584]])
tensor([[ 0.0000,  0.9379,  1.5714, -0.0693,  0.0000,  0.0000],
        [ 1.8012,  2.8181,  0.0000,  0.0000,  0.0000, -0.5107],
        [ 1.4348,  0.0000,  2.1390,  0.0000, -1.7046,  0.0000],
        [ 0.7165,  0.0000,  0.0000,  0.0000, -0.0721, -0.2628],
        [ 0.5234,  0.0077,  0.6439,  0.0000,  0.0000,  0.0000],
        [ 0.7457,  0.0000,  1.6667,  0.0000, -1.5133,  0.0000],
        [ 0.0000,  1.6385,  0.0000, -0.8872,  0.0000, -0.2752],
        [ 0.0000,  0.0000,  0.0000, -0.6760, -0.1987, -0.1372],
        [ 1.2641,  0.4682,  0.8235,  0.0000,  0.0000,  0.0000],
        [ 0.0000,  0.0000,  0.4584, -1.8230, -0.1598,  0.0000]])
```

可以看出这种方法所实现的函数功能也正确的进行了运算。这种 JIT 编译方式虽然在第一次会耗时，但是之后再次运行时则直接加载已编译好的模块。而且由于是将 C++ 代码视为字符串，其中的一些参数可以在 Python 代码中预先设置传递成为常量，从而产生针对不同参数特定编译的模块。

以上的实现都是基于 PyTorch 中 ATen 库提供的高层操作接口，实现了在 CPU 和 GPU 上共享一份代码实现。而如果想要进一步定制自己的需求，则需要利用 C++ 和 CUDA 混合编程，从底层完全自主实现，这样操作自由度更大，可以实现的代码并行度更高，能够起到计算过程加速的目的。

为了使得 Python 可以调用 CUDA kernel 函数，需要先编写 C++ 文件，调用 CUDA 实现，并绑定到 Python 中。这与之前所实现的 ncrelu.cpp 类似。可以编写如下代码在 ncrelu_cuda.cpp 中：

```cpp
#include <torch/extension.h>

// CUDA 函数声明
at::Tensor NCReLUForwardLauncher(const at::Tensor& src,
                                 const int batch,
                                 const int channels,
                                 const int height,
                                 const int width);

// 宏定义
#define CHECK_CUDA(x) AT_CHECK(x.type().is_cuda(), #x, " must be a CUDAtensor ")
#define CHECK_CONTIGUOUS(x) \
  AT_CHECK(x.is_contiguous(), #x, " must be contiguous ")
#define CHECK_INPUT(x) \
  CHECK_CUDA(x);       \
  CHECK_CONTIGUOUS(x)

// C++函数包装
at::Tensor ncrelu_forward_cuda(const at::Tensor input) {
```

```
  CHECK_INPUT(input);
  at::DeviceGuard guard(input.device());
  int batch = input.size(0);
  int channels = input.size(1);
  int height = input.size(2);
  int width = input.size(3);
  return NCReLUForwardLauncher(input, batch, channels, height, width);
}

// 绑定
PYBIND11_MODULE(TORCH_EXTENSION_NAME, m) {
  m.def("ncrelu_forward_cuda", &ncrelu_forward_cuda,
        "ncrelu forward (CUDA)");
}
```

这里第一部分是 CUDA 函数声明，将在 ncrelu_cuda_kernel.cu 文件中具体实现，稍后介绍。第二部分是定义一些宏（macro），用来辅助检测输入是否在 GPU 上，是否为连续排列存储。第三部分是 C++ 函数包装，也是最后编译出的动态链接库可调用的函数。这里的实现也很直观，获取输入的维度，并传入到 CUDA 函数中，返回得到的结果。

为了使得能够正确编译，同样需要编写 setup.py 文件。与之前实现不同的是，这里需要导入 PyTorch 中预设的 CUDAExtension 类，负责编译 C++ 和 CUDA 文件。内容如下：

```
from setuptools import setup
from torch.utils.cpp_extension import BuildExtension, CUDAExtension

setup(
    name='NCReLU',
    ext_modules=[
        CUDAExtension('ncrelu_cuda', [
            'ncrelu_cuda.cpp',
            'ncrelu_cuda_kernel.cu',
        ]),
    ],
    cmdclass={
        'build_ext': BuildExtension
    })
```

这里注意具体实现的 CUDA kernel 文件后缀名为.cu，并且其名称需要与 C++ 文件名不同。在运行 setup.py 时，会使用 nvcc 针对.cu 文件进行编译，并最终包含进动态链接库中。

最后是 ncrelu_cuda_kernel.cu 的具体实现。涉及到 CUDA 编程模型及相关知识，具体可参阅 NVIDIA 官网教程[①]，这里只作一些基本概念介绍。该文件大致分为如下结构：

---

① https://devblogs.nvidia.com/even-easier-introduction-cuda

```cpp
#include <ATen/ATen.h>
#include <ATen/cuda/CUDAContext.h>
#include <THC/THCAtomics.cuh>

template <typename scalar_t>
__global__ void NCReLUForward(/*some args*/) {
  /* implementations */
}

at::Tensor NCReLUForwardLauncher(/*some args*/) {
  /* input transformation, data structure initialization */
  /* pass to CUDA function, return output */
}
```

可以看到分成两部分。首先是一个 CUDA 特有声明为 __global__ 的模板函数，是具体执行运算的部分。其次是一个启动 CUDA kernel 的函数，这里负责对输入进行一些转换，数据初始化，并最终调用具体执行运算部分的函数，返回结果。第一部分的具体实现为：

```cpp
template <typename scalar_t>
__global__ void NCReLUForward(const int input_size,
                              const int channels,
                              const int height,
                              const int width,
                              const scalar_t * src_data,
                              scalar_t * dst_data) {

  const int index = blockIdx.x * blockDim.x + threadIdx.x;

  if (index > input_size) return;
  auto value = src_data[index];
  const int chw = channels * height * width;
  dst_data[index + index / chw * chw] = value >= 0 ? value : scalar_t(0);
  dst_data[index + index / chw * chw + chw] = value >= 0 ? scalar_t(0) : value;
}
```

该函数传递输入大小、通道数、长度、宽度、源数据指针、返回数据指针这六个参数。这里 blockIdx、blockDim、threadIdx 分别表示 block 索引、block 维度、thread 索引。这涉及到 CUDA 的编程模型。可以简单地理解为在 GPU 上有多个并发的线程同时负责以上计算，这些线程在 CUDA 编程模型里又被分为 thread、block、grid 等层次组织。index=blockIdx.x * blockDim.x + threadIdx.x 这一语句用来计算一个绝对索引，负责返回数据中某个位置处值的计算。这样只需要关注于单个线程计算过程，不需要串行的循环，因此达到了加速效果。而之后的代码也很直观，就是根据这个索引，寻找到原数据值，判断是否大于 0，并且在通道维度，按照前一半为正，后一半为负填充结果。

而第二部分启动 CUDA kernel 的函数具体实现如下：

```cpp
at::Tensor NCReLUForwardLauncher(const at::Tensor& src,
                                const int batch,
                                const int channels,
                                const int height,
                                const int width) {
    at::Tensor dst = at::empty({batch, 2 * channels, height, width},
                    src.options());
    const int input_size = batch * channels * height * width;
    const int output_size = batch * channels * height * width;
    AT_DISPATCH_FLOATING_TYPES_AND_HALF(src.scalar_type(),
"NCReLUForwardLauncher", ([&] {
        const scalar_t *src_ = src.data<scalar_t>();
        scalar_t *dst_ = dst.data<scalar_t>();

        NCReLUForward<scalar_t>
            <<<GET_BLOCKS(output_size), THREADS_PER_BLOCK,
            0, at::cuda::getCurrentCUDAStream()>>>(
                input_size, channels, height, width, src_, dst_
            );
    }));
    THCudaCheck(cudaGetLastError());
    return dst;
}
```

这里函数的输入参数为 ATen Tensor 类型的源数据，以及数据的四个维度。然后先去开辟一段存储空间，负责承接输出的计算结果。这里也是用到了 at::empty 高层接口操作，再计算输入大小。接下来是最关键的部分，利用到了 AT_DISPATCH_FLOATING_TYPES 这个宏，实现了动态分发机制（dynamic dispatch），即它会在运行时，根据输入具体的数值类型，去决定之前 CUDA kernel 模块函数需要实例化为哪种函数。这个宏需要传入数据类型，用来报错的字符串信息和一个 lambda 函数。其中数据类型由 src.scalar_type() 获取，lambda 函数是后面调用 CUDA kernel 函数部分。这其中在从 ATen Tensor 中获取某一类型数据指针后，用到了<<< >>>这一写法启动 kernel。其中需要根据输出大小分配 block 数，并设置每一 block 中的 thread 数。函数实现为：

```cpp
#define THREADS_PER_BLOCK 1024

inline int GET_BLOCKS(const int N) {
    int optimal_block_num = (N + THREADS_PER_BLOCK - 1) / THREADS_PER_BLOCK;
    int max_block_num = 65000;
    return min(optimal_block_num, max_block_num);
}
```

最后在调用了 CUDA kernel 计算结果之后，进行最后的检查，如果无报错则返回结果。

将以上代码放在同一文件夹下，并且运行 setup.py 进行编译，会出现以下信息，显示整体编译过程。

```
running build_ext
building 'ncrelu_cuda' extension
gcc -pthread -Wsign-compare -DNDEBUG -g -fwrapv
  -O3 -Wall -Wstrict-prototypes -fPIC
  -I/PATH/TO/torch/include
  -I/PATH/TO/torch/include/torch/csrc/api/include
  -I/PATH/TO/torch/include/TH
  -I/PATH/TO/torch/include/THC
  -I/usr/local/cuda/include
  -I/PATH/TO/include/python3.5m
  -c ncrelu_cuda.cpp
  -o build/temp.linux-x86_64-3.5/ncrelu_cuda.o
  -DTORCH_API_INCLUDE_EXTENSION_H
  -DTORCH_EXTENSION_NAME=ncrelu_cuda
  -D_GLIBCXX_USE_CXX11_ABI=0 -std=c++11
/usr/local/cuda/bin/nvcc -I/PATH/TO/torch/include
  -I/PATH/TO/torch/include/torch/csrc/api/include
  -I/PATH/TO/torch/include/TH
  -I/PATH/TO/torch/include/THC
  -I/usr/local/cuda/include
  -I/PATH/TO/include/python3.5m
  -c ncrelu_cuda_kernel.cu
  -o build/temp.linux-x86_64-3.5/ncrelu_cuda_kernel.o
  -D__CUDA_NO_HALF_OPERATORS__
  -D__CUDA_NO_HALF_CONVERSIONS__
  -D__CUDA_NO_HALF2_OPERATORS__
  --compiler-options '-fPIC'
  -DTORCH_API_INCLUDE_EXTENSION_H
  -DTORCH_EXTENSION_NAME=ncrelu_cuda
  -D_GLIBCXX_USE_CXX11_ABI=0 -std=c++11
g++ -pthread -shared -L/PATH/TO/lib
  -Wl,-rpath=/PATH/TO/lib,
  --no-as-needed build/temp.linux-x86_64-3.5/ncrelu_cuda.o
  build/temp.linux-x86_64-3.5/ncrelu_cuda_kernel.o
  -L/usr/local/cuda/lib64
  -L/PATH/lib -lcudart -lpython3.5m
  -o build/lib.linux-x86_64-3.5/ncrelu_cuda.cpython-35m-x86_64-linux-gnu.so
copying build/lib.linux-x86_64-3.5/ncrelu_cuda.cpython-35m-x86_64-linux-gnu.so ->
```

最后我们再测试所实现的功能。

```
>> import torch
>> import ncrelu_cuda
>> a = torch.randn(10, 3)
>> a
tensor([[ 0.3389, -0.4630,  0.2758],
```

```
        [-0.0831,  1.2776,  1.2038],
        [-0.1535, -0.1835,  0.6132],
        [-0.5139,  0.8223,  0.1366],
        [ 0.0137,  1.1834, -0.7755],
        [ 2.7611, -0.6977,  0.1868],
        [-0.4419, -0.3151,  0.4341],
        [-0.9931,  1.3653, -0.5693],
        [-0.2414,  1.9879, -2.0690],
        [-0.0860, -1.2936,  0.3629]])
>>> a = a.view(10, 3, 1, 1).cuda()
>>> b = ncrelu_cuda.ncrelu_forward_cuda(a)
>>> b = b.squeeze()
>>> b
tensor([[ 0.3389,  0.0000,  0.2758,  0.0000, -0.4630,  0.0000],
        [ 0.0000,  1.2776,  1.2038, -0.0831,  0.0000,  0.0000],
        [ 0.0000,  0.0000,  0.6132, -0.1535, -0.1835,  0.0000],
        [ 0.0000,  0.8223,  0.1366, -0.5139,  0.0000,  0.0000],
        [ 0.0137,  1.1834,  0.0000,  0.0000,  0.0000, -0.7755],
        [ 2.7611,  0.0000,  0.1868,  0.0000, -0.6977,  0.0000],
        [ 0.0000,  0.0000,  0.4341, -0.4419, -0.3151,  0.0000],
        [ 0.0000,  1.3653,  0.0000, -0.9931,  0.0000, -0.5693],
        [ 0.0000,  1.9879,  0.0000, -0.2414,  0.0000, -2.0690],
        [ 0.0000,  0.0000,  0.3629, -0.0860, -1.2936,  0.0000]],
       device='cuda:0')
```

可以看到实现结果正确符合预期。有关更多利用 C++/CUDA 实现自定义算子的内容，我们在之后的应用实战中会进一步介绍。

### 5.2.2 利用 TorchScript 导出 PyTorch 模型

上面介绍的是利用 C++ 扩展 PyTorch，也就是最终的目的还是服务于 PyTorch 在 Python 内的接口。而在一些实际应用场景中，需要部署已训练好的模型，此时往往需要利用 C++ 语言与模型进行交互。在 PyTorch 1.2 版本之后，推出了 TorchScript 这一解决方案，用来追踪导出模型结构，并且可以直接在 C++ 中进行加载。

为了实现以上目的，TorchScript 支持两种模型导出模式，一种称为 tracing，一种称为 scripting。其中 tracing 指的是 PyTorch 在模型运行时，追踪运行经过的模块算子，实时构建计算流图，并最终总结成为一种中间表示，方便在 C++ 中加载调用。这种方法对于一般静态模型有用，但是无法追踪到模型中的动态运行结构，比如与依赖于数据的分支判断。此时需要应用 scripting，从 Python 源代码级别进行解析，而非在运行时构建。但这种方式目前只支持编译部分 Python 语法，像继承、生成器等复杂的控制流无法进行转换。

这里先介绍 tracing 方式。此处我们以 ResNet50 模型为例。

```
import torch
```

```python
from torchvision import models

model = models.resnet50(pretrained=True)
model.eval()
input = torch.rand(1, 3, 224, 224)
traced_module = torch.jit.trace(model, input)
traced_module.save('traced_resnet50.pt')
```

这里通过 torch.jit.trace 方式,即将模型与输入张量一同追踪导出。注意此处模型设置为.eval()模式后,在追踪导出的模型中也设置为测试模式,并且模型的预训练权重也一并导出,说明在进行 tracing 时,模型的属性等信息也被记录下来。我们可以加载并验证该模型运行行为与原始模型一致。

```
>> import torch
>> traced_model = torch.jit.load('traced_resnet50.pt')
>> input = torch.randn(1, 3, 224, 224)
>> traced_output = traced_model(input)
>> from torchvision import models
>> model = models.resnet50(pretrained=True).eval()
>> output = model(input)
>> torch.allclose(traced_output, output)
True
```

这一例子展示了在 Python 内可以无需知道模型结构权重等信息,直接加载追踪出的模型进行推理运行。进一步的,该追踪模型甚至还利用 PyTorch 提供的 C++前端 API 加载并运行。这里我们展示一个简单 C++应用,利用追踪好的 ResNet50 模型进行图片分类。以下是 main.cpp 中的主程序代码(需要先安装 OpenCV):

```cpp
#include <torch/script.h>                           # 引用"一站式"头文件
#include <opencv2/opencv.hpp>
#include <iostream>
#include <memory>

using namespace cv;

int main(int argc, const char* argv[]) {
  std::string image_path = "./cat.jpg";             # 待加载图片路径
  Mat image = imread(image_path);                    # 读入图片

  torch::jit::script::Module module;
  module = torch::jit::load(argv[1]);                # 加载 traced module
  torch::Tensor tensor_image = torch::from_blob(     # 将图转换为 Tensor
    image.data, {1, image.rows, image.cols, 3}, torch::kByte
  );
  tensor_image = tensor_image.permute({0, 3, 1, 2});  # NHWC --> NCHW
  tensor_image = tensor_image.toType(torch::kFloat);
  tensor_image = tensor_image.div(255);              # 归一化
```

```cpp
    std::vector<torch::jit::IValue> inputs;
    inputs.push_back(tensor_image);                         # 作为输入
    at::Tensor output = module.forward(inputs).toTensor();  # 返回输出
    at::Tensor probability = at::softmax(output, /*dim=*/1); # 生成概率
    auto results = at::topk(                                # 获取前 5 个最高预测
      probability, /*k=*/5, /*dim=*/1,
      /*largest=*/true, /*sorted=*/true
    );
    auto prob = std::get<0>(results);                       # 获取概率并输出
    std::cout << prob << '\n';
    auto index = std::get<1>(results);                      # 获取类别号并输出
    std::cout << index << '\n';
}
```

然后需要编写 CMakeLists.txt 文件以编译此段代码。内容如下：

```
cmake_minimum_required(VERSION 3.0 FATAL_ERROR)
project(resnet50-cpp)

find_package(Torch REQUIRED)
find_package(OpenCV REQUIRED)

add_executable(main main.cpp)
target_link_libraries(main ${TORCH_LIBRARIES} ${OpenCV_LIBS})
set_property(TARGET main PROPERTY CXX_STANDARD 11)
```

为了编译只应用于 CPU 上的代码，故在 PyTorch 官网上下载 LibTorch[①]，并解压到当前文件夹下得到 libtorch。然后执行以下命令：

```
$ mkdir build
$ cd build
$ cmake -DCMAKE_PREFIX_PATH=./libtorch ..
$ cmake --build . --config Release
```

如果依赖配置等正确，则会输出以下信息：

```
-- The C compiler identification is GNU 6.5.0
-- The CXX compiler identification is GNU 7.4.0
-- Check for working C compiler: /usr/bin/cc
-- Check for working C compiler: /usr/bin/cc -- works
-- Detecting C compiler ABI info
-- Detecting C compiler ABI info - done
-- Detecting C compile features
-- Detecting C compile features - done
-- Check for working CXX compiler: /usr/bin/c++
-- Check for working CXX compiler: /usr/bin/c++ -- works
-- Detecting CXX compiler ABI info
-- Detecting CXX compiler ABI info - done
```

---

① https://download.pytorch.org/libtorch/cpu/libtorch-cxx11-abi-shared-with-deps-1.2.0.zip

```
-- Detecting CXX compile features
-- Detecting CXX compile features - done
-- Looking for pthread.h
-- Looking for pthread.h - found
-- Looking for pthread_create
-- Looking for pthread_create - not found
-- Looking for pthread_create in pthreads
-- Looking for pthread_create in pthreads - not found
-- Looking for pthread_create in pthread
-- Looking for pthread_create in pthread - found
-- Found Threads: TRUE
-- Found torch: /PATH/TO/libtorch/lib/libtorch.so
-- Found OpenCV: /PATH/TO/OPENCV (found version "3.2.0")
-- Configuring done
-- Generating done
-- Build files have been written to: /PATH/TO/build
Scanning dependencies of target main
[ 50%] Building CXX object CMakeFiles/main.dir/main.cpp.o
[100%] Linking CXX executable main
[100%] Built target main
```

然后返回上一级文件夹，并放置名为的 cat.jpg 图片（如图 5.1 中左侧所示）。为了方便运算，图片已经调整为 224×224 大小。运行 ./build/main traced_resnet50.pt 则可输出以下结果：

```
 0.5509  0.1991  0.0634  0.0368  0.0285
[ Variable[CPUFloatType]{1,5} ]
 285  283  287  281  282
[ Variable[CPULongType]{1,5} ]
```

其对应的类别名显示在图 5.1 右侧中。

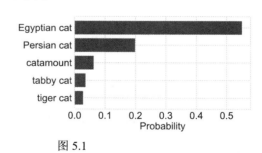

图 5.1

接下来介绍 TorchScript 支持的另一种模式 scripting。此处我们展示其如何处理模型中具有依赖输入而不同计算结构的情况（例子来源于链接[①]）。

```
import torch.nn as nn
import torch
```

---

① https://program-transformations.github.io/slides/pytorch_neurips.pdf

```python
class MyModule(nn.Module):
    def __init__(self, N, M, state):
        super(MyModule, self).__init__()
        self.weight = nn.Parameter(torch.rand(N, M))
        self.state = state

    def forward(self, input):
        self.state.append(input)
        if input.sum() > 0:
            output = self.weight.mv(input)
        else:
            output = self.weight + input
        return output

my_module = MyModule(3, 4, [torch.randn(3, 4)])
script_module = torch.jit.script(my_module)
script_module.save('script_module.pt')
```

这里构建的模块中存在根据不同输入而决定运算过程的情况。我们可以加载保存好的模型，观察其内容。

```
>> import torch
>> script_module = torch.jit.load('script_module.pt')
>> script_module.graph
graph(%self.1 : ClassType<script_module>,
      %input.1 : Tensor):
  %8 : int? = prim::Constant()
  %10 : int = prim::Constant[value=0]()
  %21 : int = prim::Constant[value=1]()
  %3 : Tensor[] = prim::GetAttr[name="state"](%self.1)
  %5 : Tensor[] = aten::append(%3, %input.1)
  %9 : Tensor = aten::sum(%input.1, %8)
  %11 : Tensor = aten::gt(%9, %10)
  %13 : bool = aten::Bool(%11)
  %output : Tensor = prim::If(%13)
    block0():
      %15 : Tensor = prim::GetAttr[name="weight"](%self.1)
      %output.1 : Tensor = aten::mv(%15, %input.1)
      -> (%output.1)
    block1():
      %19 : Tensor = prim::GetAttr[name="weight"](%self.1)
      %output.2 : Tensor = aten::add(%19, %input.1, %21)
      -> (%output.2)
  return (%output)
```

可以看出 script_module 里包含的图是一种 JIT IR（intermediate representation），并且其中有关于根据输入决定分支的情况（block0，block1）和链表附加的运算（%5 行），都被正确反映在了所构建的图中。

相比于 tracing 模式，使用 scripting 所导出的模型，还可以在构建图阶段进行优化，如常见的公共子表达式消除（common sub-expression elimination）、循环展开（loop unrolling）、点运算融合（point-wise operation fusion）等等，可以使得导出的结构更高效。这里以一个 LSTMCell 为例：

```
@torch.jit.script
def LSTMCell(x, hx, cx, w_ih, w_hh, b_ih, b_hh):
    x_mm = x.mm(w_ih.t())
    h_mm = hx.mm(w_hh.t())
    gates = x_mm + h_mm + b_ih + b_hh
    ingate, forgetgate, cellgate, outgate = gates.chunk(4, 1)
    ingate = torch.sigmoid(ingate)
    forgetgate = torch.sigmoid(forgetgate)
    cellgate = torch.tanh(cellgate)
    outgate = torch.sigmoid(outgate)
    cy = (forgetgate * cx) + (ingate * cellgate)
    hy = outgate * torch.tanh(cy)
    return hy, cy
```

这里 torch.jit.script 作为修饰器，将普通的函数修饰成为 ScriptFunction，以便于后续导出。以上 LSTMCell 过程可以看到，是对 gates 中各个部分进行逐点运算，因此可以进行融合。

```
>> import torch
>> torch._C._jit_override_can_fuse_on_cpu(True)
>> x = torch.randn(1, 3)
>> hx = torch.randn(1, 4)
>> cx = torch.randn(1, 4)
>> w_ih = torch.randn(16, 3)
>> w_hh = torch.randn(16, 4)
>> b_ih = torch.randn(16)
>> b_hh = torch.randn(16)
>> print(LSTMCell.graph_for(x, hx, cx, w_ih, w_hh, b_ih, b_hh))
graph(%x.1 : Float(*, *),
      %hx.1 : Float(*, *),
      %cx.1 : Float(*, *),
      %w_ih.1 : Float(*, *),
      %w_hh.1 : Float(*, *),
      %b_ih.1 : Float(*),
      %b_hh.1 : Float(*)):
  %9 : Float(*, *) = aten::t(%w_ih.1)
  %x_mm.1 : Float(*, *) = aten::mm(%x.1, %9)
  %11 : Float(*, *) = aten::t(%w_hh.1)
  %h_mm.1 : Float(*, *) = aten::mm(%hx.1, %11)
  %77 : Tensor[] = prim::ListConstruct(%b_hh.1, %b_ih.1, %x_mm.1, %h_mm.1)
  %78 : Tensor[] = aten::broadcast_tensors(%77)
  %79 : Tensor,%80:Tensor,%81:Tensor,%82:Tensor = prim::ListUnpack(%78)
  %hy.1 : Float(*,*),%cy.1:Float(*,*)=
```

```
            prim::FusionGroup_0(%cx.1,%82,%81,%80,%79)
  %30 : (Float(*, *), Float(*, *)) = prim::TupleConstruct(%hy.1, %cy.1)
  return (%30)
```

可以看出在设置为算子可融合模式后,输出的 JIT IR 中出现了 FusionGroup_0 算子。而该算子即为融合后的算子。

拥有了以上的 JIT IR,则可以进一步根据不同平台和硬件翻译成实际底层运算代码。这里我们展示如何将 BatchNorm 和 ReLU 运算利用 scripting 导出的 JIT IR 构建融合后的 C++算子实现。首先导出 BatchNorm 和 ReLU 操作的中间表示图。

```
import torch

@torch.jit.script
def BNReLU(x, mean, var, alpha, gamma):
    std = torch.sqrt(var)
    x = x - mean
    x = x / (std + 1e-05)
    x = x * alpha
    x = x + gamma
    x = torch.relu(x)
    return x

x = torch.randn(2, 4, 5, 5)
mean = x.mean(dim=(0,2,3), keepdim=True)
var = x.var(dim=(0,2,3), keepdim=True)
alpha = torch.randn(1, 4, 1, 1)
gamma = torch.randn(1, 4, 1, 1)
inputs = [x, mean, var, alpha, gamma]

graph = BNReLU.graph_for(*inputs)
```

这里我们可以展示所生成的 IR:

```
graph(%x.1 : Float(*, *, *, *),
      %mean.1 : Float(*, *, *, *),
      %var.1 : Float(*, *, *, *),
      %alpha.1 : Float(*, *, *, *),
      %gamma.1 : Float(*, *, *, *)):
  %5 : int = prim::Constant[value=1]()
  %6 : float = prim::Constant[value=1e-05]()
  %std.1 : Float(*, *, *, *) = aten::sqrt(%var.1)
  %x.3 : Float(*, *, *, *) = aten::sub(%x.1, %mean.1, %5)
  %9 : Float(*, *, *, *) = aten::add(%std.1, %6, %5)
  %x.5 : Float(*, *, *, *) = aten::div(%x.3, %9)
  %x.7 : Float(*, *, *, *) = aten::mul(%x.5, %alpha.1)
  %x.9 : Float(*, *, *, *) = aten::add(%x.7, %gamma.1, %5)
  %x.11 : Float(*, *, *, *) = aten::relu(%x.9)
  return (%x.11)
```

然后利用其融合机制生成 CPU kernel 代码：

```
>> torch._C._jit_override_can_fuse_on_cpu(True)
>> code = torch._C._jit_fuser_get_fused_kernel_code(graph, inputs)
>> print(code)
```

输出的代码为：

```
#include <math.h>
#include <cstddef>
#include <cstdint>

double rsqrt(double x) {
  return 1.0/sqrt(x);
}

float rsqrtf(float x) {
  return 1.0f/sqrtf(x);
}

double frac(double x) {
  return x - trunc(x);
}

float fracf(float x) {
  return x - truncf(x);
}

#define POS_INFINITY INFINITY
#define NEG_INFINITY -INFINITY

typedef unsigned int IndexType;
template<typename T, size_t N>
struct TensorInfo {
  T* data;
  IndexType sizes[N];
  IndexType strides[N];
};
template<typename T>
struct TensorInfo<T, 0> {
  T * data;
};


#ifdef _MSC_VER
template<size_t n> struct int_of_size;

#define DEFINE_INT_OF_SIZE(int_t) \
template<> struct int_of_size<sizeof(int_t)> { using type = int_t; }
```

```cpp
DEFINE_INT_OF_SIZE(int64_t);
DEFINE_INT_OF_SIZE(int32_t);
DEFINE_INT_OF_SIZE(int16_t);
DEFINE_INT_OF_SIZE(int8_t);

#undef DEFINE_INT_OF_SIZE

template <typename T>
using int_same_size_t = typename int_of_size<sizeof(T)>::type;

#define IndexTypeLoop int_same_size_t<IndexType>
#define ToIndexTypeLoop(x) static_cast<IndexTypeLoop>(x)
#else
#define IndexTypeLoop IndexType
#define ToIndexTypeLoop(x) x
#endif

#define OMP_THRESHOLD 100000
static void kernel_0_kernel(
    IndexType totalElements, const TensorInfo<float,1> t0,
    const TensorInfo<float,3> t1, const TensorInfo<float,3> t2,
    const TensorInfo<float,3> t3, const TensorInfo<float,3> t4,
    const TensorInfo<float,1> t5) {
  #pragma omp parallel for if(totalElements > OMP_THRESHOLD)
  for (IndexTypeLoop linearIndex = 0;
       linearIndex < ToIndexTypeLoop(totalElements);
       linearIndex += 1) {
    // Convert `linearIndex` into an offset of tensor:
    IndexType t0_offset = 0;
    IndexType t0_linearIndex = linearIndex;

    size_t t0_dimIndex0 = t0_linearIndex ;
    t0_offset += t0_dimIndex0 ;
    IndexType t1_offset = 0;
    IndexType t1_linearIndex = linearIndex;
    ...
    ...
    // calculate the results
    float n0 = t0.data[t0_offset];
    float n1 = t1.data[t1_offset];
    float n2 = t2.data[t2_offset];
    float n3 = t3.data[t3_offset];
    float n4 = t4.data[t4_offset];
    int64_t n5 = 1;
    double n6 = 1e-05;
    float n7 = sqrtf(n2);
    float n8 = (n0 - ((float) n5)*n1);
    float n9 = n7 + ((float) n5)*((float) n6);
```

```
        float n10 = n8 / n9;
        float n11 = n10 * n3;
        float n12 = n11 + ((float) n5)*n4;
        float n13 = n12 < 0 ? 0.f : n12 ;
        t5.data[t5_offset] = n13;
    }
}

#ifdef _WIN32
#define JIT_API __declspec(dllexport)
#else
#define JIT_API
#endif

extern "C"
JIT_API void kernel_0(IndexType totalElements, void ** args) {
  kernel_0_kernel(
    totalElements ,
    *static_cast<TensorInfo<float,1>*>(args[1]),
    *static_cast<TensorInfo<float,3>*>(args[2]),
    *static_cast<TensorInfo<float,3>*>(args[3]),
    *static_cast<TensorInfo<float,3>*>(args[4]),
    *static_cast<TensorInfo<float,3>*>(args[5]),
    *static_cast<TensorInfo<float,1>*>(args[6])
  );
}
```

为了节省空间，这里将中间较为重复的部分略去。可以看出以上通过 IR 转换自动生成依赖于硬件和平台代码十分方便。当前关于 TorchScript 的开发与研究还处于活跃期，一些功能和接口也会在不断变动。更多文档信息和教程可参考官方链接[1][2]。

## 5.3 丰富的 PyTorch 资源介绍

最后介绍一些 PyTorch 学习资源。首先最主要的 PyTorch 资料索引和使用介绍就是其官方文档[3]及教程[4]。这是在使用 PyTorch 中遇到问题查询的最直接途径。其官方教程中也包含了多种任务及使用情形展示，非常易于理解。这些文档教程也被翻译成了中文版[5]，供国内读者学习。而如果想进一步深入 PyTorch 源码内部进行理解，参考其 github 上的开源代码库[6]，其中不仅包括了 PyTorch 核心源码，还有像 torchvision、torchtext，以及官

---

[1] https://pytorch.org/blog/optimizing-cuda-rnn-with-torchscript/
[2] https://pytorch.org/docs/stable/jit.html
[3] https://pytorch.org/docs/stable/index.html
[4] https://pytorch.org/tutorials/
[5] https://pytorch.apachecn.org
[6] https://github.com/pytorch

方示例 example 等。PyTorch 还搭建了一个讨论平台[①]，供使用者提问讨论，很多 PyTorch 核心开发者都会及时进行回复。而 github issue 则更多是服务于开发环节提问，避免请教基本使用问题。

除此以外，便是多关注基于 PyTorch 实现的工程及论文相关代码。如对于物体检测领域比较著名的有港中文开发的 mmdetection[②]、Facebook 官方实现的 maskrcnn-benchmark[③]。对于图像分类除了 torchvision 内置的 models 模块，当前追踪新型模型较全面的应属 pytorch-image-models 代码库[④]。对于自然语言处理则有 Allen 研究院推出的 AllenNLP[⑤]，还有最近流行的自然语言预训练模型库 Transformers[⑥]。对于语音识别则有 pytorch-kaldi 工具箱[⑦]，对于概率编程则有 Uber AI Lab 推出的 pyro[⑧] 等等。可以看出诸多领域均有基于 PyTorch 实现的框架及代码库，不一而足。这里推荐一个总结 Awesome-PyTorch-List[⑨]，供供读者尽情探索。

## 5.4 PyTorch 应用实战一：在 ImageNet 上训练 MobileNet-V2 网络

本节中我们将综合运用之前所介绍的 PyTorch 并行计算技巧，在 ImageNet 数据集上从头开始完整训练 MobileNet-V2 网络。ImageNet 数据集总体大约包含 120 万张训练图片和 5 万张验证集图片，总大小在 140GB 左右，需要在 ImageNet 官网链接下载[⑩]。在下载完之后需要解压，并且根据类别将其归类为文件夹。其中 train 文件下已经归类完毕，共 1000 类。val 文件夹下需要利用单独脚本[⑪] 进行分类组织。接下来我们就可以开始整体训练实验。

### 5.4.1 MobileNet-V2 网络介绍

首先介绍 MobileNet-V2 网络结构。该网络提出于论文 "MobilenetV2: Inverted Residuals and Linear Bottlenecks"[48]之中。该文章主要提出了一种新的模块结构 Inverted Residual Block，如图 5.2（b）所示（取自原文），类似于 ResNet 中的 Residual Block，该

---

① https://discuss.pytorch.org
② https://github.com/open-mmlab/mmdetection
③ https://github.com/facebookresearch/maskrcnn-benchmark
④ https://github.com/rwightman/pytorch-image-models
⑤ https://github.com/allenai/allennlp
⑥ https://github.com/huggingface/transformers
⑦ https://github.com/mravanelli/pytorch-kaldi
⑧ https://github.com/pyro-ppl/pyro
⑨ https://github.com/bharathgs/Awesome-pytorch-list
⑩ http://image-net.org
⑪ https://raw.githubusercontent.com/soumith/imagenetloader.torch/master/valprep.sh

模块中也存在着两路分支结构。重点区别在于计算残差分支的构造。ResNet 中的残差部分是将中间 feature map 通道数减小，而在 Inverted 残差模块中则是通过 1×1 卷积将其扩张，然后后续连接 Depthwise 卷积，最后再通过 1×1 卷积缩小通道数目。而且在最后的残差部分与跨层连接相加部分无 ReLU 函数激活输出，以防止破坏特征的表达能力。

MobileNet-V2 整体结构如图 5.2 中右图所示，其中 $t$ 指的是 Inverted Residual Block 中间特征通道数目扩展倍数，$c$ 指的是该层输入特征通道数目，$n$ 为模块重复个数，$s$ 为该部分第一个模块卷积 stride 数目。MobileNet-V2 通过利用 Depthwise 卷积可以有效降低计算量，减少占用内存数量，同时且可以取得不错的模型预测效果。torchvision 中提供预训练好的 mobilenet_v2 模型及权重，在 ImageNet 上可以达到 71.88%的 top-1 准确率和 90.29%的 top-5 准确率。我们之后的实现就是去达到这一水平。

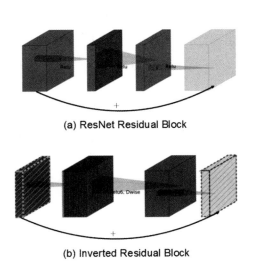

图 5.2

## 5.4.2 具体实现

首先我们封装加载数据集的部分在文件 dataset.py 中。这里应用到了在 5.1.1 节中介绍的 fast_collate_fn 和 DataPrefetcher 技巧。而在数据增广部分遵循常见的实践技巧[①]，添加了 ColorJitter、Lightning、RandomHorizontalFlip 操作。

```
import torch
import numpy as np
from PIL import Image
import torchvision.transforms as tfm
from torchvision.datasets import ImageFolder
```

---

① https://github.com/eladhoffer/convNet.pytorch/blob/master/preprocess.py

```python
import torch.utils.data as data

imagenet_pca = {
    'eigval': np.asarray([0.2175, 0.0188, 0.0045]),
    'eigvec': np.asarray([
        [-0.5675, 0.7192, 0.4009],
        [-0.5808, -0.0045, -0.8140],
        [-0.5836, -0.6948, 0.4203],
    ])
}

class Lighting(object):
    def __init__(self, alphastd,
                 eigval=imagenet_pca['eigval'],
                 eigvec=imagenet_pca['eigvec']):
        self.alphastd = alphastd
        assert eigval.shape == (3,)
        assert eigvec.shape == (3, 3)
        self.eigval = eigval
        self.eigvec = eigvec

    def __call__(self, img):
        if self.alphastd == 0.:
            return img
        rnd = np.random.randn(3) * self.alphastd
        rnd = rnd.astype('float32')
        v = rnd
        old_dtype = np.asarray(img).dtype
        v = v * self.eigval
        v = v.reshape((3, 1))
        inc = np.dot(self.eigvec, v).reshape((3,))
        img = np.add(img, inc)
        if old_dtype == np.uint8:
            img = np.clip(img, 0, 255)
        img = Image.fromarray(img.astype(old_dtype), 'RGB')
        return img

    def __repr__(self):
        return self.__class__.__name__ + '()'


def fast_collate(batch):
    imgs = [img[0] for img in batch]
    targets = torch.tensor(
        [target[1] for target in batch], dtype=torch.int64
    )
    w = imgs[0].size[0]
    h = imgs[0].size[1]
```

```python
        tensor = torch.zeros((len(imgs), 3, h, w), dtype=torch.uint8)
        for i, img in enumerate(imgs):
            nump_array = np.asarray(img, dtype=np.uint8)
            if(nump_array.ndim < 3):
                nump_array = np.expand_dims(nump_array, axis=-1)
            nump_array = np.rollaxis(nump_array, 2)
            tensor[i] += torch.from_numpy(nump_array)

        return tensor, targets


def get_imagenet_loader(root, batch_size, type='train'):
    crop_scale = 0.25
    jitter_param = 0.4
    lighting_param = 0.1
    if type == 'train':
        transform = tfm.Compose([
            tfm.RandomResizedCrop(224, scale=(crop_scale, 1.0)),
            tfm.ColorJitter(
                brightness=jitter_param, contrast=jitter_param,
                saturation=jitter_param),
            Lighting(lighting_param),
            tfm.RandomHorizontalFlip(),
        ])

    elif type == 'test':
        transform = tfm.Compose([
            tfm.Resize(256),
            tfm.CenterCrop(224),
        ])

    dataset = ImageFolder(root, transform)
    sampler = data.distributed.DistributedSampler(dataset)
    data_loader = data.DataLoader(
        dataset, batch_size=batch_size,
        shuffle=False, num_workers=4,
        pin_memory=True, sampler=sampler,
        collate_fn=fast_collate
    )
    if type == 'train':
        return data_loader, sampler

    elif type == 'test':
        return data_loader


class DataPrefetcher():
    def __init__(self, loader):
```

```
        self.loader = iter(loader)
        self.stream = torch.cuda.Stream()
        self.mean = torch.tensor(
            [0.485 * 255, 0.456 * 255, 0.406 * 255]
        ).cuda().view(1,3,1,1)
        self.std = torch.tensor(
            [0.229 * 255, 0.224 * 255, 0.225 * 255]
        ).cuda().view(1,3,1,1)
        self.preload()

    def preload(self):
        try:
            self.next_input, self.next_target = next(self.loader)
        except StopIteration:
            self.next_input = None
            self.next_target = None
            return
        with torch.cuda.stream(self.stream):
            self.next_input = self.next_input.cuda(async=True)
            self.next_target = self.next_target.cuda(async=True)
            self.next_input = self.next_input.float()
            self.next_input = self.next_input.sub_(self.mean).div_(self.std)

    def next(self):
        torch.cuda.current_stream().wait_stream(self.stream)
        input = self.next_input
        target = self.next_target
        self.preload()
        return input, target
```

然后便是核心训练代码 main.py。这里我们采用 apex 提供的 amp 模块和 DistributedDataParallel，实现单机多卡分布式混合精度训练。优化中使用 SGD 优化器和 CosineAnnealing 学习率调整器进行训练，并且对于所有的 Depthwise 卷积权重不添加 weight decay 操作。在设置批样本量和学习率时应注意，此时的 batch size 指的是每个进程加载的 batch size，因此实际上每次迭代经过的总批样本量大小应为 world_size × batch_size。而根据论文[50]所提出的 linear scaling rule，当 batch size 扩大 $k$ 倍时，初始学习率也应扩大 $k$ 倍，才可以使得训练水平达到小批样本量下的情况。因此我们以 batch_size=256 时，lr=0.05 为基础而等比例调整学习率。

在训练过程中，还增加 label smoothing[49]操作作为额外正则化，提高模型泛化能力。label smoothing 具体操作即对计算交叉熵损失函数时，使用的 one-hot 形式标签，转变为包含其他类别少量信息的浮点值标签。例如原始四分类标签为[0, 1, 0, 0]，现在变为[0.1, 0.7, 0.1, 0.1]，其表达式为（为$K$总类别数目，$\alpha$为平滑加权比例）：

$$y^{LS} = (1-\alpha) * y^{onehot} + \alpha/K$$

main.py 具体代码如下（改编自 apex 示例[①]）：

```python
import torch
import argparse
import os
import torch.distributed as dist
import torch.nn as nn
import torch.optim as optim
import dataset
import misc
from apex.parallel import DistributedDataParallel as DDP
from torchvision import models
from apex.fp16_utils import *
from apex import amp

parser = argparse.ArgumentParser()
parser.add_argument('--data', default='imagenet', type=str)
parser.add_argument('--arch', '-a', default='mobilenet_v2', type=str)
parser.add_argument('--lr', default=0.05, type=float)
parser.add_argument('--mm', default=0.9, type=float)
parser.add_argument('--wd', default=4e-5, type=float)
parser.add_argument('--epochs', default=150, type=int)
parser.add_argument('--log_interval', default=50, type=int)
parser.add_argument('-b', '--batch_size', default=256, type=int,
            help='mini-batch size per process (default: 256)')
parser.add_argument("--local_rank", default=0, type=int)
parser.add_argument('--label_smooth', action='store_true')
parser.add_argument('--opt-level', type=str)
parser.add_argument('--keep-batchnorm-fp32', type=str, default=None)
parser.add_argument('--loss-scale', type=str, default=None)

args = parser.parse_args()
args.logdir = 'baseline-%s-fp16' % (args.arch)
if args.label_smooth:
    args.logdir += '-labelsmooth'

# 使用 cudnn 加速
torch.backends.cudnn.benchmark = True

# 设置进程对应 GPU
args.gpu = args.local_rank % torch.cuda.device_count()
torch.cuda.set_device(args.gpu)
torch.distributed.init_process_group(backend='nccl',
                            init_method='env://')
args.world_size = torch.distributed.get_world_size()
```

---

[①] https://github.com/NVIDIA/apex/blob/master/examples/imagenet/main_amp.py

```python
    if args.local_rank == 0:
        misc.prepare_logging(args)

    # 只让 rank 编号为 0 的进程输出训练信息
    def print(msg):
        if args.local_rank == 0:
            misc.logger.info(msg)

    # 获取多进程间平均的张量值
    def reduce_tensor(tensor):
        rt = tensor.clone()
        dist.all_reduce(rt, op=dist.reduce_op.SUM)
        rt /= args.world_size
        return rt

    def to_python_float(t):
        if hasattr(t, 'item'):
            return t.item()
        else:
            return t[0]

print("=> Using model {}".format(args.arch))
model = models.mobilenet_v2()
model = model.cuda()

criterion = nn.CrossEntropyLoss().cuda()

# 使用 label_smooth
if args.label_smooth:
    class CrossEntropyLabelSmooth(nn.Module):
        def __init__(self, num_classes, epsilon):
            super(CrossEntropyLabelSmooth, self).__init__()
            self.num_classes = num_classes
            self.epsilon = epsilon
            self.logsoftmax = nn.LogSoftmax(dim=1)

        def forward(self, inputs, targets):
            log_probs = self.logsoftmax(inputs)
            targets = torch.zeros_like(log_probs).scatter_(
                1, targets.unsqueeze(1), 1
            )
            targets = (1 - self.epsilon) * targets \
                + self.epsilon / self.num_classes
            loss = (-targets * log_probs).mean(0).sum()
            return loss

    criterion = CrossEntropyLabelSmooth(
        num_classes=1000, epsilon=0.1
```

```python
    ).cuda()

# 根据总批样本量大小等比例调整学习率
args.lr = args.lr*float(args.batch_size*args.world_size)/256.

# 所有 Depthwise 卷积权重 (N, 1, x, x) 无 weight decay
model_params = []
for params in model.parameters():
    ps = list(params.size())
    if len(ps) == 4 and ps[1] != 1:
        weight_decay = args.wd
    elif len(ps) == 2:
        weight_decay = args.wd
    else:
        weight_decay = 0
    item = {'params': params, 'weight_decay': weight_decay,
            'lr': args.lr, 'momentum': args.mm,
            'nesterov': True}
    model_params.append(item)

optimizer = torch.optim.SGD(model_params)
lr_scheduler = optim.lr_scheduler.CosineAnnealingLR(
    optimizer, T_max=args.epochs, eta_min=0
)

# 利用 amp 进行封装
model, optimizer = amp.initialize(model, optimizer,
                                  opt_level=args.opt_level,
                                  keep_batchnorm_fp32=args.keep_batchnorm_fp32,
                                  loss_scale=args.loss_scale
                                  )
# 使用 DistributedDataParallel 进行分布式训练
model = DDP(model)

print('==> Preparing data..')
train_loader, train_sampler = dataset.get_imagenet_loader(
    os.path.join(args.data, 'train'), args.batch_size, type='train'
)
test_loader = dataset.get_imagenet_loader(
    os.path.join(args.data, 'val'), 100, type='test'
)

# 求平均值的记录器
class AverageMeter(object):
    def __init__(self):
        self.reset()

    def reset(self):
```

```python
            self.val = 0
            self.avg = 0
            self.sum = 0
            self.count = 0

        def update(self, val, n=1):
            self.val = val
            self.sum += val * n
            self.count += n
            self.avg = self.sum / self.count

    def accuracy(output, target, topk=(1,)):
        maxk = max(topk)
        batch_size = target.size(0)

        _, pred = output.topk(maxk, 1, True, True)
        pred = pred.t()
        correct = pred.eq(target.view(1, -1).expand_as(pred))

        res = []
        for k in topk:
            correct_k = correct[:k].view(-1).float().sum(0, keepdim=True)
            res.append(correct_k.mul_(100.0 / batch_size))
        return res

    def train(train_loader, model, criterion, optimizer, epoch):
        losses = AverageMeter()
        top1 = AverageMeter()
        top5 = AverageMeter()

        # 利用 DataPrefetcher 迭代数据
        prefetcher = dataset.DataPrefetcher(train_loader)
        model.train()

        input, target = prefetcher.next()
        i = -1
        while input is not None:
            i += 1

            output = model(input)
            loss = criterion(output, target)

            # 计算梯度并更新权重
            optimizer.zero_grad()
            with amp.scale_loss(loss, optimizer) as scaled_loss:
                scaled_loss.backward()
            optimizer.step()
```

```python
        if i % args.log_interval == 0:
            prec1, prec5 = accuracy(output.data, target, topk=(1, 5))

            # 求平均loss和准确率，需使用 reduce_tensor
            reduced_loss = reduce_tensor(loss.data)
            prec1 = reduce_tensor(prec1)
            prec5 = reduce_tensor(prec5)

            losses.update(to_python_float(reduced_loss), input.size(0))
            top1.update(to_python_float(prec1), input.size(0))
            top5.update(to_python_float(prec5), input.size(0))

            print('Epoch: [{0}][{1}/{2}]\t'
                  'Loss {loss.val:.10f} ({loss.avg:.4f})\t'
                  'Prec@1 {top1.val:.3f} ({top1.avg:.3f})\t'
                  'Prec@5 {top5.val:.3f} ({top5.avg:.3f})'.format(
                   epoch, i, len(train_loader),
                   loss=losses, top1=top1, top5=top5
            ))

        input, target = prefetcher.next()


def validate(val_loader, model, criterion, epoch):
    losses = AverageMeter()
    top1 = AverageMeter()
    top5 = AverageMeter()

    prefetcher = dataset.DataPrefetcher(val_loader)

    # 转换为eval模式
    model.eval()

    input, target = prefetcher.next()
    i = -1
    while input is not None:
        i += 1
        with torch.no_grad():
            output = model(input)
            loss = criterion(output, target)

        # measure accuracy and record loss
        prec1, prec5 = accuracy(output.data, target, topk=(1, 5))

        reduced_loss = reduce_tensor(loss.data)
        prec1 = reduce_tensor(prec1)
        prec5 = reduce_tensor(prec5)
```

```
        losses.update(to_python_float(reduced_loss), input.size(0))
        top1.update(to_python_float(prec1), input.size(0))
        top5.update(to_python_float(prec5), input.size(0))

        input, target = prefetcher.next()

    print(' * Test Epoch {0}, Prec@1 {top1.avg:.3f} '
          'Prec@5 {top5.avg:.3f}\n'
            .format(epoch, top1=top1, top5=top5))


for epoch in range(args.epochs):
    # 设置 sampler 内部 epoch 值
    # 使得每一周期随机种子不同，打乱重排顺序
    train_sampler.set_epoch(epoch)
    train(train_loader, model, criterion, optimizer, epoch)
    validate(test_loader, model, criterion, epoch)

    # 保存模型及优化器状态
    if args.local_rank == 0:
        torch.save({
            'epoch': epoch + 1,
            'arch': args.arch,
            'state_dict': model.state_dict(),
            'optimizer' : optimizer.state_dict(),
        }, os.path.join(args.logdir, 'checkpoint.pth'))

    # 调整学习率
    lr_scheduler.step()
```

最后运行如下命令：

```
CUDA_VISIBLE_DEVICES=0,1,2,3 \                  # 指定 GPU 卡号
 python -m torch.distributed.launch \           # 启动分布式进程
        --nproc_per_node=4 main.py \            # 使用四个进程
        -a mobilenet_v2 -b 320 \                # batch_size=320
        --opt-level O1 --loss-scale 128.0       # 混合精度训练参数
```

这里我们使用了 4 个 NVIDIA GeForce GTX 1080 Ti 显卡，显存最大为 11GB，以上的 batch size 设置恰好可使得显存填满，并且 GPU 利用率可以达到 99%以上，说明此时数据加载速度够快，没有让 GPU 计算资源闲置。最后经过训练可以达到 71.994%的 top-1 准确率和 90.608%的准确率，比 torchvision 的预训练模型准确率还要稍高一些。图 5.3 也展示了用 32 位全精度训练和混合精度训练的损失函数值及准确率对比情况。可以看出二者相差无几，说明了采用混合精度训练可以保证无性能损失。

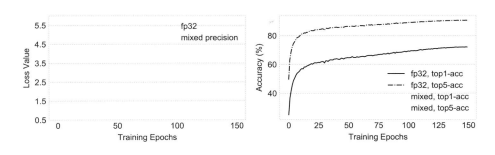

图 5.3

## 5.5 PyTorch 应用实战二：利用 CUDA 扩展实现 MixConv 算子

之前在介绍使用 C++和 CUDA 混合编程实现自定义算子时，所举得例子颇为简单，造成了一种"多此一举"的感觉。在这一节中，我们将展示利用底层 CUDA kernel 实现的 MixConv 算子，相比于在 Python 中利用 PyTorch 的高层接口实现，可以起到真实加速的效果。这不仅可以显著减少模型推理时延，也提供了一种新的算子高层接口，方便开发更多的模型和算法。

### 5.5.1 MixConv 算子介绍

这里我们先介绍 MixConv 算子具体构造。该算子首次提出于论文"MixConv: Mixed Depthwise Convolution Kernels"[51]中。其核心思想是改造传统的 Depthwise 卷积，使其中可以允许多种不同卷积核存在，按照不同通道输出。传统的 Depthwise 卷积如图 5.4 中（a）所示（取自原论文），对于输入的 $C$ 个通道，每个通道都进行二维的卷积，卷积核为 $k \times k$，输出和输入有着相同的通道数目。这样操作可以极大地减少卷积过程中的运算量。而文章所提出的 MixConv 则是考虑在其中不同的通道上使用不同的卷积核进行运算。文中观察到用这种混合卷积方式代替原模型中的 Depthwise 卷积部分之后，在物体检测和图像分类任务上都能提升性能，并且由于其依然使用了类似 Depthwise 卷积方式，计算量相比于标准卷积还是有极大的减少。

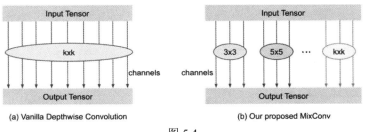

图 5.4

而对于这个 MixConv，其实现方式也是比较直观。代码如下[①]：

```python
import torch
import torch.nn as nn

def _split_channels(num_chan, num_groups):      # 根据组数对输入通道分组
    split = [num_chan // num_groups for _ in range(num_groups)]
    split[0] += num_chan - sum(split)
    return split

class MixConv(nn.Module):
    def __init__(self,                           # 默认 dilation 均为 1
                 in_channels,
                 kernel_size=[3,5,7],
                 padding=[1,2,3],
                 stride=1):
        super(MixConv, self).__init__()
        if padding is None:                      # 默认为 same padding 模式
            padding = [(k-1)//2 for k in kernel_size]

        self.num_groups = len(kernel_size)
        self.in_splits = _split_channels(in_channels, self.num_groups)
        self.layer = nn.ModuleList([])           # 按照每个分组初始化卷积
        for c, k, p in zip(self.in_splits, kernel_size, padding):
            self.layer.append(
                nn.Conv2d(c, c, k, stride, p, groups=c, bias=False)
            )

    def forward(self, x):
        out = []
        x_split = torch.split(x, self.in_splits, dim=1)
        for m, _x in zip(self.layer, x_split):   # 循环计算每个分组输出
            out.append(m(_x))

        return torch.cat(out, dim=1)             # 按通道维度拼接在一起
```

以上的实现虽然简单直观，但是在前向计算中，需要涉及循环计算，这其实破坏了原本 Depthwise 卷积计算所具有的并行度。所以以上的实现在实际运行时反而会增加延时并未启动加速效果。因此如何消除掉 Python 层面的循环调度，从 CUDA kernel 底层上依然实现各通道并行计算，是我们改造的主要目的。

### 5.5.2 借鉴 Depthwise 卷积实现思路

在实现 MixConv 之前，我们先观察 Depthwise 卷积实现的思路。这里列举出一个直

---

[①] 部分代码借鉴自 https://github.com/rwightman/pytorch-image-models/blob/master/timm/models/layers/mixed_conv2d.py

观实现的核心代码：

```
1  CUDA_1D_KERNEL_LOOP(index, output_size) {
2    const int n = index / channels / height / width;
3    const int c = (index / height / width) % channels;
4    const int h = (index / width) % height;
5    const int w = index % width;
6    const scalar_t *weight = weight_data + c * kernel_h * kernel_w;
7    scalar_t value = 0;
8    for (int kh = 0; kh < kernel_h; ++kh) {
9      for (int kw = 0; kw < kernel_w; ++kw) {
10       const int h_in = -pad_h + h * stride_h + kh * dilation_h;
11       const int w_in = -pad_w + w * stride_w + kw * dilation_w;
12       if ((h_in >= 0) && (h_in < bottom_height)
13         && (w_in >=0) && (w_in < bottom_width)) {
14         const int offset = ((n * channels + c) * bottom_height + h_in)
15           * bottom_width + w_in;
16         value += (*weight) * bottom_data[offset];
17       }
18       ++weight;
19     }
20   }
21   top_data[index] = value;
22 }
```

这里第 1 行 CUDA_1D_KERNEL_LOOP 是一个常用的宏定义，就是之前我们所介绍的用 thread 索引、block 索引等信息计算出全局索引的方式，来模拟一种循环。但是实际上各线程是独立并行的。接下来 2~5 行代码是通过此全局索引，计算出该位置元素位于输出张量中所在的批样本数位置、通道位置、长和宽位置。记住，这里的输入、权重以及输出张量均为一维向量。所计算出的这些指标只是为了我们理解容易，而实际上并不是以此去获取元素的，还是需要用全局索引 index 才可以。第 6 行代码是获取针对于第 c 个通道输出，权重指针应该移动到这个通道权重开始的位置处。因为已经经过了 c 个通道，且每个通道权重大小为 kernel_h×kernel_w，所以移动这么多偏移即可。此处就是之后我们实现 MixConv 时应当注意区别的地方，因为 Depthwise 卷积是默认所有卷积核大小相同的，而 MixConv 里的卷积核不同所以偏移计算方式不一样。

剩下的部分便是在 kernel_h×kernel_w 区域上，获取输入元素位置（第 14 行），然后和权重逐元素相乘并累加（第 16 行），与此同时权重指针不断移动（第 18 行），并在此区域循环结束后，将结果赋值到输出的相应位置处（第 21 行）。所以经过以上分析，我们可以理解了 Depthwise 卷积的实现过程，并且意识到如何处理权重指针位置是实现 MixConv 的关键。

### 5.5.3 具体实现

经过以上的分析,我们便可以借鉴 Depthwise 卷积实现思路,通过改造权重指针计算方式,来实现 MixConv。这里为了简化问题,我们默认 MixConv 中各个分组的卷积有着相同的 stride,并且 dilation 为 1。同时所有的卷积的 pad 会根据各自卷积核大小补零到相同尺寸。

因此,我们可以实现 mix_conv2d_kernel.cu 文件如下:

```cpp
#include <ATen/ATen.h>
#include <ATen/cuda/CUDAContext.h>
#include <THC/THCAtomics.cuh>

// 宏定义
#define CUDA_1D_KERNEL_LOOP(i, n)                           \
  for (int i = blockIdx.x * blockDim.x + threadIdx.x; i < n; \
       i += blockDim.x * gridDim.x)

#define THREADS_PER_BLOCK 1024

inline int GET_BLOCKS(const int N) {
  int optimal_block_num = (N + THREADS_PER_BLOCK - 1) / THREADS_PER_BLOCK;
  int max_block_num = 65000;
  return min(optimal_block_num, max_block_num);
}

// CUDA kernel 模板函数
template <typename scalar_t>
__global__ void MixConv2DForward(const int output_size, const int channels,
                    const int height, const int width,
                    const int64_t *weight_ptr_list,
                    const int64_t *kernel_h_list,
                    const int64_t *kernel_w_list,
                    const int64_t *pad_h_list,
                    const int64_t *pad_w_list,
                    const int stride_h, const int stride_w,
                    const int bottom_height, const int bottom_width,
                    const scalar_t *bottom_data,
                    const scalar_t *weight_data,
                    scalar_t *top_data) {
  CUDA_1D_KERNEL_LOOP(index, output_size) {
    const int n = index / channels / height / width;
    const int c = (index / height / width) % channels;
    const int h = (index / width) % height;
    const int w = index % width;

    const scalar_t *weight = weight_data + weight_ptr_list[c];
```

```cpp
    const int kernel_h = kernel_h_list[c];
    const int kernel_w = kernel_w_list[c];
    const int pad_h = pad_h_list[c];
    const int pad_w = pad_w_list[c];

    scalar_t value = 0;
    for (int kh = 0; kh < kernel_h; ++kh) {
      for (int kw = 0; kw < kernel_w; ++kw) {
        const int h_in = -pad_h + h * stride_h + kh;
        const int w_in = -pad_w + w * stride_w + kw;
        if ((h_in >= 0) && (h_in < bottom_height)
          && (w_in >=0) && (w_in < bottom_width)) {
          const int offset = ((n * channels + c) * bottom_height + h_in)
            * bottom_width + w_in;
          value += (*weight) * bottom_data[offset];
        }
        ++weight;
      }
    }
    top_data[index] = value;
  }
}

// 启动 kernel 函数
at::Tensor MixConv2DForwardLaucher(const int batch, const int channels,
                    const int height, const int width,
                    const at::Tensor weight_ptr_list,
                    const at::Tensor kernel_h_list,
                    const at::Tensor kernel_w_list,
                    const at::Tensor pad_h_list,
                    const at::Tensor pad_w_list,
                    const int stride_h, const int stride_w,
                    const at::Tensor& bottom_data,
                    const at::Tensor& weight_data) {

  const int output_h = (height - 1) / stride_h + 1;
  const int output_w = (width - 1) / stride_w + 1;
  const int output_size = batch * channels * output_h * output_w;
  at::Tensor top_data = at::empty({batch, channels, output_h, output_w},
                    bottom_data.options());

  AT_DISPATCH_FLOATING_TYPES_AND_HALF(
    bottom_data.scalar_type(), "MixConv2DLaucherForward", ([&] {
      const int64_t *weight_ptr_list_ = weight_ptr_list.data<int64_t>();
      const int64_t *kernel_h_list_ = kernel_h_list.data<int64_t>();
      const int64_t *kernel_w_list_ = kernel_w_list.data<int64_t>();
      const int64_t *pad_h_list_ = pad_h_list.data<int64_t>();
      const int64_t *pad_w_list_ = pad_w_list.data<int64_t>();
```

```cpp
            const scalar_t *bottom_data_ = bottom_data.data<scalar_t>();
            const scalar_t *weight_data_ = weight_data.data<scalar_t>();
            scalar_t *top_data_ = top_data.data<scalar_t>();

            MixConv2DForward<scalar_t>
                <<<GET_BLOCKS(output_size), THREADS_PER_BLOCK,
                0, at::cuda::getCurrentCUDAStream()>>>(
                    output_size, channels, output_h, output_w,
                    weight_ptr_list_, kernel_h_list_, kernel_w_list_,
                    pad_h_list_, pad_w_list_,
                    stride_h, stride_w, height, width,
                    bottom_data_, weight_data_, top_data_
                );
        }));
    THCudaCheck(cudaGetLastError());
    return top_data;
}
```

首先第一部分是宏定义以及一些辅助函数,我们在之前已作过讲解。第二部分是关键,实现了 CUDA kernel 模板函数。其参数列表与 Depthwise 卷积稍有不同,此处不仅传入了权重,还需要传入一个整数型 weight_ptr_list 数组,用来记录每个通道对应的卷积核权重偏移量,以便于移动权重指针。比如有 4 个通道,其卷积核大小分别为[3, 5, 7, 9],则 weight_ptr_list = [0, $3^2$, $3^2+5^2$, $3^2+5^2+7^2$] = [0, 9, 34, 83]。这样在获取每个通道权重开始位置处,便不再是用 weight_data + c * kernel_h * kernel_w 计算,而是 weight_data + weight_ptr_list[c]。传递参数中还有 kernel_h_list 这样四个整数数组,用来指定每个通道权重的卷积核大小,以及对应补零值个数。因此像指定卷积核大小操作,原来 Depthwise 卷积是作为常量值参数传入,而现在通过 kernel_h = kernel_h_list[c]这样的方式获取。之后的逐点相乘累加操作与原来相同。第三部分则是启动 CUDA kernel 函数的部分,先去计算出输出大小,并利用 ATen 封装的高层接口 at::empty 分配输出存储空间。然后便是之前介绍过的动态分发机制,调用实际的 CUDA kernel 函数,最后返回结果。

需要编写另一个文件 mix_conv2d_cuda.cpp 用来绑定到 Python 中,具体内容如下:

```cpp
#include <torch/extension.h>

#include <cmath>
#include <vector>

at::Tensor MixConv2DForwardLaucher(const int batch, const int channels,
                    const int height, const int width,
                    const at::Tensor weight_ptr_list,
                    const at::Tensor kernel_h_list,
                    const at::Tensor kernel_w_list,
```

```cpp
                      const at::Tensor pad_h_list,
                      const at::Tensor pad_w_list,
                      const int stride_h, const int stride_w,
                      const at::Tensor& bottom_data,
                      const at::Tensor& weight_data);

#define CHECK_CUDA(x) AT_CHECK(x.type().is_cuda(), #x, " must be a CUDAtensor ")
#define CHECK_CONTIGUOUS(x) \
  AT_CHECK(x.is_contiguous(), #x, " must be contiguous ")
#define CHECK_INPUT(x) \
  CHECK_CUDA(x);       \
  CHECK_CONTIGUOUS(x)

at::Tensor mix_conv2d_forward_cuda(
                      const at::Tensor im,
                      const at::Tensor weight,
                      const at::Tensor weight_ptr_list,
                      const at::Tensor kernel_h_list,
                      const at::Tensor kernel_w_list,
                      const at::Tensor pad_h_list,
                      const at::Tensor pad_w_list,
                      const int stride_h, const int stride_w) {
  CHECK_INPUT(im);
  CHECK_INPUT(weight);
  CHECK_INPUT(weight_ptr_list);
  CHECK_INPUT(kernel_h_list);
  CHECK_INPUT(kernel_w_list);
  CHECK_INPUT(pad_h_list);
  CHECK_INPUT(pad_w_list);

  at::DeviceGuard guard(im.device());
  int batch = im.size(0);
  int channels = im.size(1);
  int height = im.size(2);
  int width = im.size(3);

  return MixConv2DForwardLaucher(
    batch, channels, height, width,
    weight_ptr_list,
    kernel_h_list, kernel_w_list,
    pad_h_list, pad_w_list,
    stride_h, stride_w,
    im, weight
  );
}

PYBIND11_MODULE(TORCH_EXTENSION_NAME, m) {
  m.def("mix_conv2d_forward", &mix_conv2d_forward_cuda,
```

```
            "mixed conv2d forward (CUDA)");
}
```

其内容与之前介绍过的类似，即 mix_conv2d_forward_cuda 接收输入等参数，经过输入检查，获取大小等信息后，调用 MixConv2DForwardLauncher，完成实际的运算。以上两个文件放在 src 文件夹下，然后编写 setup.py，用以编译 C++和 CUDA 文件。

```python
from setuptools import setup
from torch.utils.cpp_extension import BuildExtension, CUDAExtension

setup(
    name='MixConv2d',
    ext_modules=[
        CUDAExtension('mix_conv2d_cuda', [
            'src/mix_conv2d_cuda.cpp',
            'src/mix_conv2d_kernel.cu',
        ]),
    ],
    cmdclass={
        'build_ext': BuildExtension
    })
```

最后运行 python setup.py build_ext --inplace 之后，如果无报错信息等，则会得到类似于 mix_conv2d_cuda.cpython-35m-x86_64-linux-gnu.so 这样的共享对象文件（中间的平台名称与所编译环境有关）。

最后便是在 PyTorch 中进行封装，使其可以成为一个 nn.Module 子类。首先是将 mix_conv2d_forward 封装成为 torch.autograd.Function 类型。

```python
import torch
from mix_conv2d_cuda import mix_conv2d_forward
from torch.autograd import Function
from torch.autograd.function import once_differentiable

class MixConv2dFunc(Function):

    @staticmethod
    def forward(ctx,
                input, weight_vector,
                weight_ptr_list,
                kernel_h_list, kernel_w_list,
                pad_h_list, pad_w_list,
                stride_h, stride_w):

        return mix_conv2d_forward(
            input, weight_vector,
            weight_ptr_list,
            kernel_h_list, kernel_w_list,
            pad_h_list, pad_w_list,
```

```
            stride_h, stride_w
        )

    @staticmethod
    @once_differentiable
    def backward(ctx, grad_output):
        return (None,) * 7
```

这里应注意，由于我们只实现了 MixConv2d 的前向运算过程，没有实现其反向求导函数，因此封装为 Function 时，backward 方法设置为全部返回 None 对象。而且即使用 CUDA kernel 实现了其反向求导函数，但是该反向函数的导数依然没有实现，因此用 once_differentiable 修饰器注明此处不支持高阶求导过程。

然后是利用 MixConv2dFunc 实现 MixConv2d Module，代码如下：

```
import torch
import torch.nn as nn
from torch.nn.utils.convert_parameters import parameters_to_vector
from torch.nn.modules.utils import _pair

# 根据输入通道，及分组个数进行分组
def _split_channels(num_chan, num_groups):
    split = [num_chan // num_groups for _ in range(num_groups)]
    split[0] += num_chan - sum(split)
    return split


class MixConv2d(nn.Module):
    def __init__(self, in_channels, kernels,
                 paddings=None, stride=(1,1)):
        super(MixConv2d, self).__init__()
        kernels = [_pair(k) for k in kernels]      # kernels存储(k, k)格式
        if paddings is None:                        # 默认是same padding模式
            paddings = [
                ((kh-1)//2, (kw-1)//2) for kh, kw in kernels
            ]
        else:
            paddings = [_pair(p) for p in paddings]

        assert len(kernels) == len(paddings)
        self.num_groups = len(kernels)
        in_splits = _split_channels(in_channels, self.num_groups)
        kernel_h_list = []
        kernel_w_list = []
        pad_h_list = []
        pad_w_list = []
        for i in range(self.num_groups):
            c = in_splits[i]
            kh, kw = kernels[i]
```

```python
            ph, pw = paddings[i]
            self.register_parameter(                  # 注册不同权重
                'weight_%dx%d' % (kh, kw),
                nn.Parameter(torch.Tensor(c, 1, kh, kw))
            )
            kernel_h_list.extend([kh] * c)
            kernel_w_list.extend([kw] * c)
            pad_h_list.extend([ph] * c)
            pad_w_list.extend([pw] * c)

        weight_ptr_list = [0]                         # 计算 weight_ptr_list
        start = 0
        for i in range(1, in_channels):
            start += kernel_w_list[i-1]*kernel_h_list[i-1]
            weight_ptr_list.append(start)

        # 以下数组虽不是可训练参数,但是需要注册在 buffer 中,加载保存模块时使用
        self.register_buffer(
            'kernel_h_list', torch.LongTensor(kernel_h_list)
        )
        self.register_buffer(
            'kernel_w_list', torch.LongTensor(kernel_w_list)
        )
        self.register_buffer(
            'pad_h_list', torch.LongTensor(pad_h_list)
        )
        self.register_buffer(
            'pad_w_list', torch.LongTensor(pad_w_list)
        )
        self.register_buffer(
            'weight_ptr_list', torch.LongTensor(weight_ptr_list)
        )
        self.stride = stride

    # 将所有参数拼接成一维向量
    def _vectorize_parameters(self):
        self.weight_vector = parameters_to_vector(self.parameters())

    # 初始化参数
    def _init_weights(self):
        for p in self.parameters():
            nn.init.kaiming_normal_(p, nonlinearity='relu')

    # 调用 Function 计算
    def forward(self, input):
        return MixConv2dFunc.apply(
            input, self.weight_vector, self.weight_ptr_list,
            self.kernel_h_list, self.kernel_w_list,
```

```
                  self.pad_h_list, self.pad_w_list,
                  self.stride[0], self.stride[1])
```

在实现了以上的 MixConv2d 模块后，我们可以和利用 PyTorch 普通实现方式的结果进行比较。

```
>> import torch
>> from mix_conv2d import MixConv2d          # CUDA 实现
>> from mixed_conv2d import MixConv          # PyTorch 实现
>> input = torch.randn(128, 480, 28, 28).cuda()
>> m1 = MixConv2d(480, [3,5,7,9])
>> m1._init_weights()                         # 初始化参数
>> m2 = MixConv(480, [3,5,7,9])
>> m2.layer[0].weight.data = m1.weight_3x3.data  # 对应权重一一赋值
>> m2.layer[1].weight.data = m1.weight_5x5.data
>> m2.layer[2].weight.data = m1.weight_7x7.data
>> m2.layer[3].weight.data = m1.weight_9x9.data
>> m1 = m1.cuda()
>> m1._vectorize_parameters()                 # 要先.cuda()后向量化参数
>> m2 = m2.cuda()
>> o1 = m1(input)
>> o2 = m2(input)
>> torch.allclose(o1, o2)                     # 二者数值近似相同
True
```

可以看到 CUDA 实现结果与 PyTorch 普通实现结果相同，证明了其正确性。最后我们再测量二者运算时的时间消耗，此处采取形状大小为[16, 512, 32, 32]的张量作为输入，然后设置 $n$ 个通道分组，每组卷积核大小为 $2n+1$，即[3, 5, 7...]，输入 padding 形式为 same padding。然后在将输入权重等张量初始化并迁移至 GPU 后，重复运行 100 次，统计平均延时。测试代码如下：

```
from mixed_conv2d import MixConv
from mix_conv2d import MixConv2d
import torch
import timeit
import os
import argparse
parser = argparse.ArgumentParser()
parser.add_argument(
    '--n_groups', '-n', default=1, type=int    # 指定分组数
)
args = parser.parse_args()

os.environ['CUDA_VISIBLE_DEVICES'] = '0'       # 指定 GPU 卡号

input = torch.randn(16,512,32,32).cuda()

kernels = []                                    # 设置每组卷积核大小
```

```python
start = 3
for i in range(args.n_groups):
    kernels.append(start)
    start += 2

m1 = MixConv(512, kernels).cuda()                 # 这里只测延时
m2 = MixConv2d(512, kernels)                      # 无需保证二者权重对应
m2._init_weights()
m2 = m2.cuda()
m2._vectorize_parameters()

def test1():                                      # 测试 GPU 上延时
    torch.cuda.synchronize()                      # 需要 synchronize 同步
    m1(input)
    torch.cuda.synchronize()

def test2():
    torch.cuda.synchronize()
    m2(input)
    torch.cuda.synchronize()

time1 = timeit.timeit(test1, number=100)          # 统计平均延时（单位 ms）
print('native_impl: %.4fms' % (time1*1000))
time2 = timeit.timeit(test2, number=100)
print('my_impl: %.4fms' % (time2*1000))
```

以上代码在 NVIDIA GeForce GTX 1080 Ti 显卡上进行测试，并统计了分组数从 1 到 10 变化时，二者计算平均延时的情况。图 5.5 展示了结果，可以看出利用 CUDA kernel 实现的 MixConv 在各种分组情况下，推理延时均小于基于 PyTorch 普通实现的模块。而且随着分组数增多，缩减延时程度更大。从这个例子可以看出掌握了底层 C++ 和 CUDA 混合编程技巧，实现自定义算子扩展 PyTorch，是在科研和工程实践中都非常有益的能力。

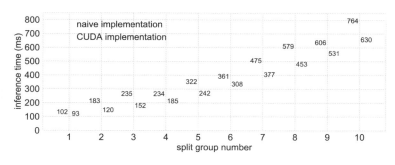

图 5.5

# 第 6 章 PyTorch 完整实战讲解——网络剪枝应用

在这一章,我们将通过一个完整的实战流程讲解,全面展示 PyTorch 在实际科研探究过程中的使用情况。此次我们选择的是网络剪枝(Network Pruning)这一领域。网络剪枝顾名思义,就是将大网络的冗余结构和参数删除,同时依然保持和大网络相近可比的预测性能,因此在一些计算密集应用场景中,通过对网络进行剪枝,可以更好地部署实际生产环境中减少计算延时和代价,具有广泛的应用前景。本章中就将带领大家熟悉了解几种经典的剪枝算法,并且在一个较为统一的框架实现,并最终进行对比评测。

## 6.1 网络剪枝介绍

本节将介绍网络剪枝的背景和概况。这里我们主要关注一大类利用权重通道重要性的结构化剪枝方法。

### 6.1.1 剪枝方法分类

首先网络剪枝这一领域研究的出现,其研究动机主要解决当前深度学习模型参数量大,计算代价高,训练时间长的问题。其研究出发点在于观察到常见的深度学习模型中都存在着冗余结构和参数,将其中大部分删除之后模型依然可以有着一定预测性能。因此如何发掘出这些潜在的高效剪枝网络(Pruned Network)结构,以及如何训练剪枝网络达到和原完整模型可比的预测性能是当前的研究重点。

剪枝技术可以先从动态剪枝(Dynamic Pruning)和静态剪枝(Static Pruning)分类。所谓动态剪枝是指模型在运行时根据不同输入情况选择实际的运行单元结构,从而起到实时的加速目的。但是这种方法在训练和测试时均需要包含完整的全模型,同时动态选择引入的额外计算负担会影响实际加速性能。而静态剪枝则是当前更为常见且实际应用更广的剪枝方法,其主要结果就是寻找一个固定的子网络结构,不同输入运行时均共享同一模型。这种寻找子网络结构的思路其实和当前研究火热的神经架构搜索(Neural Architecture Search,NAS)领域中的一些工作异曲同工,都是希望从冗余的结构中挖掘出高效子网络。

对于静态剪枝,可以进一步分为非结构化剪枝(Unstructured Pruning)和结构化剪枝(Structured Pruning)。这二者的区分主要强调于剪除单元的粒度(Granularity)上。前者往往是对权重各个元素的剪枝,从而形成了稀疏张量。而后者往往在权重通道维度、分支结构等更大层面上进行删除。因此非结构化剪枝往往可以取得更高的稀疏度[52][53],但是需要特定的软硬件支持才可以取得加速效果[54]。而结构化剪枝虽然无法做到极限的剪枝压缩程度，但是由于保留下的子结构规整，可以像完整模型一样直接在 GPU 上运行张量计算操作，更容易取得加速效果，拥有广泛的应用场景。而我们此处实战主要研究的即是结构化剪枝算法。接下来我们将介绍这一类算法的解决方式。

### 6.1.2 基于权重通道重要性的结构化剪枝

这里我们介绍一大类主要的结构化剪枝算法，即基于权重通道重要性（Weight Channel Importance）为依据的剪枝算法。其思想很直观，通过各种方式，求得当前冗余模型中各个权重通道的重要性，然后排序将重要性更高的权重通道保留下来，便得到剪枝子网络。这种基于权重通道粒度的剪枝算法也被称为 Channel Pruning。这里主要介绍三种方法，L1-norm[55]剪枝、Network Slimming[56]和 Soft Filter Pruning[57]。

首先 L1-norm 剪枝是最简单直观的，求权重通道重要性即根据各个权重的 L1 范数决定，更大的范数意味着更高的重要性。而这种方法需要事先获取预训练好的模型。

然后是 Network Slimming 方法。从这里开始，权重通道的重要性需要根据数据学习得到，因此相当于额外增加一组可训练参数，在训练过程中除了学习原始完整模型的权重，还需要学习这一组重要性值。为了促使这组重要性稀疏（Sparse），优化学习时除了常规的分类损失函数外，还需要添加这组重要性值的 L1 范数，以此作为正则项。待学习完重要性值之后，再根据其绝对值大小排序，保留最重要的结构。

最后是 Soft Filter Pruning 方法。也类似于 Network Slimming，在训练阶段同时学习通道重要性值。但是 Network Slimming 只在完整学习结束后进行一次剪枝删除。Soft Filter Pruning 则是在训练过程中每次迭代时就进行剪枝。但是又不是 Hard Pruning，只是将对于权重通道的重要性值置为 0，起到抑制输出为 0 的作用。而在之后的迭代学习过程中，又允许其再次更新，因此为 Soft Pruning。在完整学习通道重要性结束后，再进行一次剪枝得到最终子结构。

除了这种基于学习通道重要性剪枝的策略之外，还有更多结构化剪枝算法，如基于层内输出重构最小误差或类似于 LASSO[58]问题进行权重压缩的策略[59][60]，基于强化学习问题建模学习得到逐层通道重要性的策略[61]等等。此处所选择实现的三种算法建模问题方式较为统一，故重点论述研究。

### 6.1.3 问题定义与建模

为了学习得到权重通道的重要性，这里采用在权重对应输出的特征图上添加控制门（Control Gate）方式进行建模。具体来说，对于卷积层每个输出通道上，附属一个标量值控制门 $\lambda$，在卷积层完成得到输出后，再将所有控制门逐通道（Channel-Wise）地与输出特征进行相乘。如果 $\lambda$ 接近于 0，则说明该层特征可以被完全删除而不会影响最终模型预测结果，也即对应的权重通道可以被剪枝删去。图 6.1 展示了这个过程。

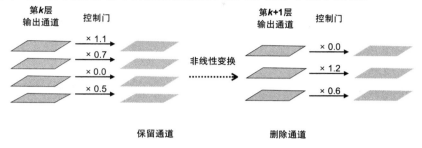

图 6.1

具体来说，对于模型中共 $L$ 层待剪枝的模块，所有控制门记作 $\Lambda = \{\lambda_1, \lambda_2, \cdots, \lambda_L\}$（注意这里 $\lambda_i$ 为一组控制门值构成的向量），则总体优化目标为：

$$\min_{\Lambda} \sum_{i}^{N} \mathcal{L}(f_\theta(x_i; \Lambda), y_i) + \gamma \cdot \Omega(\Lambda)$$

其中 $\{x_i, y_i\}$ 为数据样本及对应标签类别，$\mathcal{L}$ 为交叉熵损失函数，$\theta$ 为模型权重参数，$\Omega$ 为正则项促使控制门值尽量稀疏，$\gamma$ 为调节正则项强度的系数。在优化时，模型权重参数也一起进行学习，可以更好地适应控制门的调节。

## 6.2 具体实现思路

由上面的讨论中我们可以看出，得到权重通道重要性值的关键即是学习附属的控制门值。那么如何在网络中的计算模块附属控制门值是实现难点。其次在得到重要性值之后，如何进一步剪枝得到子结构也是需要考虑的问题。最后得到剪枝子网络之后如何进一步训练使其可以达到和完整模型可比的预测性能也是需要考虑的重点。本节将对这三方面进行一一讲解。

### 6.2.1 如何附属控制门值

在学习控制门之前，需要将网络中待剪枝模块单元内部附属上控制门值。如果仅从为了学习控制门角度来看，可以直接在编写模块单元代码时，额外添加这样一组控制门可训练参数，并在 nn.Module 的 Forward 方法中，利用控制门值逐通道与输出特征相乘。但是这样会造成两个不便之处，第一就是侵入式地修改了原网络内部实现，第二就是在剪枝结束后，训练子网络时，前向计算中不再需要控制门的调节了，需要恢复成为原始普通输出过程。如果重新编写普通模块则会造成代码重复。

在这里我们展示利用 Python 中的 Types 库实现动态替换前向输出过程。这里我们以一个简单示例展示其用法：

```
>> import types
>> class A(object):                                    # 类的定义
>>     def forward(self, x):
>>         return x
>>
>> def new_forward(self, x):                           # 待替换的函数方法
>>     return x * 2
>>
>> a = A()                                             # 实例化类 A
>> a.forward = types.MethodType(new_forward, a)       # 动态绑定到 a 的 forward
>> a.forward(1)                                        # 实际调用 new_forward 函数
2
```

从以上过程可以看出，types.MethodType 实现了在实例化类后，动态绑定一个函数到该实例的方法上。这样我们就可以先像往常搭建正常模型一样书写，然后在初始化模型后，再根据替换单元的类型，绑定利用控制门调节输出的新前向计算过程。而在剪枝结束后，重新根据剪枝结构初始化模型，则依然使用了正常的前向计算函数，减少了代码重复。具体的实现我们将在 6.3.1 节中详细介绍。

所附属的控制门位置也和待剪枝的计算模块单元类型有关。一般来说在简单的 Conv-BN-ReLU 结构中，我们在 BN 的后面按照输出通道个数附属控制门变量，即可起到删减冗余输出通道的作用。而对于带有跨层连接的模块如 ResNet 中的 Residual Block，此时应只在残差分支上附属控制门。比如 BasicBlock 的残差分支上有两个卷积层，仅在第一个卷积层的 BN 之后附属控制门变量。这样可以使得中间的特征转换维度下降，而残差分支的输入和输出通道依然保持一致，方便跨层连接相加。

### 6.2.2 剪枝结构搜索

在训练完各层的控制门值后，如何获得剪枝结构是下一步解决的问题。在原始的 Network Slimming 和 Soft Filter Pruning 论文中，常用的做法是直接按照控制门总数一定比例进行删减。这样的做法会忽视了不同层计算量的不同。更合理的做法应该是都删减模型至某一计算量约束之下。这里衡量模型计算量方法我们使用的是常用的 FLOPS 指标，通过计算浮点数操作量，衡量网络中如卷积层、池化层的乘累加量（MAC）。这里推荐使用一个第三方实现 ptflops 库，通过安装 pip install --upgrade git+https://github.com/sovrasov/flops-counter.pytorch.git，即可全局使用。使用方法也很简单：

```
>>> from torchvision import models
>>> from ptflops import get_model_complexity_info
>>> m = models.alexnet()
>>> macs, params = get_model_complexity_info(m, (3, 224, 224))
```

```
AlexNet(
  61.101 M, 100.000% Params, 0.716 GMac, 100.000% MACs,
  (features): Sequential(
    2.47 M, 4.042% Params, 0.657 GMac, 91.806% MACs,
    (0): Conv2d(0.023 M, 0.038% Params, 0.07 GMac, 9.849% MACs, 3, 64,
      kernel_size=(11, 11), stride=(4, 4), padding=(2, 2))
    (1): ReLU(0.0 M, 0.000% Params, 0.0 GMac, 0.027% MACs, inplace=True)
    (2): MaxPool2d(0.0 M, 0.000% Params, 0.0 GMac, 0.027% MACs,
      kernel_size=3, stride=2, padding=0, dilation=1, ceil_mode=False)
    (3): Conv2d(0.307 M, 0.503% Params, 0.224 GMac, 31.317% MACs, 64,
      192, kernel_size=(5, 5), stride=(1, 1), padding=(2, 2))
    (4): ReLU(0.0 M, 0.000% Params, 0.0 GMac, 0.020% MACs, inplace=True)
    (5): MaxPool2d(0.0 M, 0.000% Params, 0.0 GMac, 0.020% MACs,
      kernel_size=3, stride=2, padding=0, dilation=1, ceil_mode=False)
    (6): Conv2d(0.664 M, 1.087% Params, 0.112 GMac, 15.681% MACs, 192,
      384, kernel_size=(3, 3), stride=(1, 1), padding=(1, 1))
    (7): ReLU(0.0 M, 0.000% Params, 0.0 GMac, 0.009% MACs, inplace=True)
    (8): Conv2d(0.885 M, 1.448% Params, 0.15 GMac, 20.902% MACs, 384,
      256, kernel_size=(3, 3), stride=(1, 1), padding=(1, 1))
    (9): ReLU(0.0 M, 0.000% Params, 0.0 GMac, 0.006% MACs, inplace=True)
    (10): Conv2d(0.59 M, 0.966% Params, 0.1 GMac, 13.937% MACs, 256, 256,
      kernel_size=(3, 3), stride=(1, 1), padding=(1, 1))
    (11): ReLU(0.0 M, 0.000% Params, 0.0 GMac, 0.006% MACs, inplace=True)
    (12): MaxPool2d(0.0 M, 0.000% Params, 0.0 GMac, 0.006% MACs,
      kernel_size=3, stride=2, padding=0, dilation=1, ceil_mode=False)
  )
  (classifier): Sequential(
    58.631 M, 95.958% Params, 0.059 GMac, 8.194% MACs,
    (0): Dropout(0.0 M, 0.000% Params, 0.0 GMac, 0.000% MACs, p=0.5,
      inplace=False)
    (1): Linear(37.753 M, 61.788% Params, 0.038 GMac, 5.276% MACs,
      in_features=9216, out_features=4096, bias=True)
    (2): ReLU(0.0 M, 0.000% Params, 0.0 GMac, 0.001% MACs, inplace=True)
    (3): Dropout(0.0 M, 0.000% Params, 0.0 GMac, 0.000% MACs, p=0.5,
      inplace=False)
    (4): Linear(16.781 M, 27.465% Params, 0.017 GMac, 2.345% MACs,
      in_features=4096, out_features=4096, bias=True)
    (5): ReLU(0.0 M, 0.000% Params, 0.0 GMac, 0.001% MACs, inplace=True)
    (6): Linear(4.097 M, 6.705% Params, 0.004 GMac, 0.572% MACs,
      in_features=4096, out_features=1000, bias=True)
  )
)
>>> macs, params
('0.72 GMac', '61.1 M')
```

可以看出通过传入模型并制定模型的输入尺寸大小,即可以输出模型中各层的参数量及 MAC 数量,非常方便。我们在使用时可以设置不输出各层信息,并且令返回值为浮点数而不是字符串。

有了以上方便计算模型计算量的工具，我们就可以设计剪枝结构搜索算法。这里所谓的搜索就是根据已经学习得到的各计算单元层的控制门值后，如何决定哪些非重要的权重通道可以被删除，哪些可以被保留下来。相比于 NAS 那种整体网络拓扑结构都需要的搜索，这里的剪枝结构搜索更加局部微观，因为只是对各层权重通道个数的搜索，是很单一的维度。这里我们采取的搜索逻辑是简单的二分法。首先决定一个剪枝比例 $r$，然后按照控制门重要性排序保留最重要的部分。最后计算该剪枝结构的计算量参数量信息。如果超过预定约束，则说明保留的过多，应该提高剪枝比例，令更多的通道被删除掉。然后通过不断的二分区间查找试错，直到所得到的最终网络结构计算量在预定约束附近，则停止搜索。整体的搜索剪枝结构算法如下。

输入：$r_{\max} = 1$，$r_{\min} = 0$，迭代次数 $T$，剪枝器 Pruner，目标计算约束 $C$，偏差容忍值 $e$
输出：剪枝模型结构 $A$

1： 对于 $t = 1, 2, \ldots, T$
2：　　　$r_{\mathrm{mid}} = (r_{\max} + r_{\min}) / 2$
3：　　　$A_t$ = Pruner($r_{\mathrm{mid}}$)　　　　　　// 根据剪枝比例得到剪枝结构
4：　　　$C_t$ = calculate_complexity($A_t$)　　// 根据当前结构计算模型复杂度
5：　　　如果 $|C_t - C| / C \le e$，则
6：　　　　　$A = A_t$
7：　　　　　直接结束
8：　　　如果 $C_t < C$，则 $r_{\max} = r_{\mathrm{mid}}$，否则 $r_{\min} = r_{\mathrm{mid}}$

### 6.2.3 剪枝模型训练

最后我们需要关注在获得剪枝网络结构之后，如何进一步训练。在传统的剪枝流程中，采取的是继承原完整模型中的权重，然后在剪枝网络结构上进一步微调。这种做法一开始被认为是很直观的。因为剪枝结构的筛选就是来源于这些原始权重，基于这部分权重进一步调整，使得其适应新的剪枝结构，以便于得到更高的预测性能。然而在实际训练中与这一想法差距很大。因为很多时候删减过后的模型，在尚未进行微调时，预测性能会产生巨大的下降。而此时进行的微调，对原始模型权重已经产生很大的改变。所以这一步权重继承并没有必然合理性。

实际上在最近的研究中已经发现了这个问题。在论文"Rethinking the Value of Network Pruning"[62]中作者就指出，其实剪枝模型在得到剪枝结构后，只需要从完全随机初始化权重开始重新训练，完全可以达到与原模型可相比的预测性能。这实际上就排除了权重继承这一步的作用。这种训练方式还要求训练时长增加。在原论文中称这样的训练计划

称为 Scratch-B，"B"的代表了 Budget，意思是要求剪枝模型在整体训练过程中的计算代价要和原完整模型相同。也就是剪枝模型的计算量占原模型的比例为 $r$，则训练时长应该延长 $1/r$ 倍。这一训练过程示意图展示在图 6.2 中。

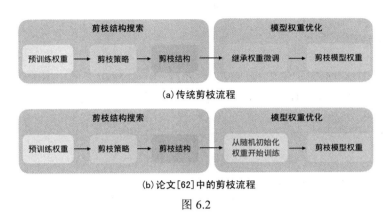

图 6.2

在后续的实验中我们会发现，这是提高剪枝模型预测性能的必要手段。

## 6.3 完整代码实现

接下来我们介绍实际完整代码的实现。整体代码主要分为三个部分，首先是训练控制门值阶段，此时通过各种剪枝方法的训练策略，得到模型中每个通道的重要性。其次是剪枝阶段，根据已经训练得到的控制门值，按照前述的剪枝结构搜索策略，得到最终的剪枝网络结构。最后是训练阶段，采取从头开始训练的 Scratch-B 训练策略得到最终模型预测准确率。本节将对各个部分进行介绍，整体工程文件结构如下：

```
torch-prune
├── data                                    # 数据集
│   └── cifar10
│       ├── cifar-10-batches-py
│       │   ├── batches.meta
│       │   ├── data_batch_1
│       │   ├── data_batch_2
│       │   ├── data_batch_3
│       │   ├── data_batch_4
│       │   ├── data_batch_5
│       │   ├── readme.html
│       │   └── test_batch
│       └── cifar-10-python.tar.gz
├── layer.py                                # 主要模型计算层
├── learn_gates.py                          # 学习控制门变量
├── misc.py                                 # 辅助函数
├── models                                  # 主要模型
│   ├── __init__.py
```

```
│       ├── resnet.py
│       └── vgg.py
├── prune_gates.py                              # 剪枝模型
├── pruner.py                                   # 剪枝器
└── train_model.py                              # 训练模型
```

### 6.3.1 模型搭建

这一部分介绍如何搭建模型。首先是 layer.py 文件。里面实现了主要模型关键层的计算过程。此处我们以 Conv-BN-ReLU 这一常见组合层为例进行展示，其代码如下：

```python
import torch
import torch.nn as nn


class ConvBNReLU(nn.Module):
    def __init__(self, in_channel, out_channel,
                 kernel_size=3, stride=1,
                 padding=1, bias=False):
        super(ConvBNReLU, self).__init__()
        self.conv = nn.Conv2d(in_channel, out_channel,
                              kernel_size, stride,
                              padding, bias=bias)
        self.bn = nn.BatchNorm2d(out_channel)
        self.relu = nn.ReLU()

    def forward(self, x):                                    # 正常前向计算
        out = self.conv(x)
        out = self.bn(out)
        out = self.relu(out)
        return out

    def init_gates(self):                                    # 初始化通道控制门变量
        self.gates = nn.Parameter(
            torch.ones(self.conv.out_channels)
        )

    def get_gates(self):                                     # 获取控制门变量
        return [self.gates]

    def gated_forward(self, *input, **kwargs):    # 控制门调节下的前向计算过程
        out = self.conv(input[1])
        out = self.bn(out)
        out = self.gates.view(1, -1, 1, 1) * out
        out = self.relu(out)
        return out
```

可以看出相比于普通的 Module，添加了有关控制门变量部分的代码，并且也实现了在控制门变量调节下的前向计算过程，之后会用来进行替换。ResNet 中的 BasicBlock 也类似这样实现：

```python
class BasicBlock(nn.Module):
    def __init__(self,
                 in_channel,
                 mid_channel,
                 out_channel,
                 kernel_size=3,
                 stride=1,
                 padding=1,
                 bias=False):
        super(BasicBlock, self).__init__()
        self.conv1 = nn.Conv2d(
            in_channel, mid_channel, kernel_size,
            stride, padding, bias=bias
        )
        self.bn1 = nn.BatchNorm2d(mid_channel)
        self.relu1 = nn.ReLU()
        self.conv2 = nn.Conv2d(
            mid_channel, out_channel, kernel_size,
            1, padding, bias=bias
        )
        self.bn2 = nn.BatchNorm2d(out_channel)
        self.relu2 = nn.ReLU()
        self.shortcut = nn.Sequential()

        if stride != 1 or out_channel != in_channel:
            self.shortcut = nn.Sequential(
                nn.Conv2d(
                    in_channel, out_channel, 1,
                    stride, 0, bias=bias
                ),
                nn.BatchNorm2d(out_channel)
            )

    def forward(self, x):
        out = self.conv1(x)
        out = self.bn1(out)
        out = self.relu1(out)
        out = self.conv2(out)
        out = self.bn2(out)
        out += self.shortcut(x)
        out = self.relu2(out)
        return out

    def init_gates(self):
```

```python
        self.gates = nn.Parameter(
            torch.ones(self.conv1.out_channels)
        )

    def get_gates(self):
        return [self.gates]

    def gated_forward(self, *input, **kwargs):
        out = self.conv1(input[1])
        out = self.bn1(out)
        out = self.gates.view(1, -1, 1, 1) * out
        out = self.relu1(out)
        out = self.conv2(out)
        out = self.bn2(out)
        out += self.shortcut(input[1])
        out = self.relu2(out)
        return out
```

以上代码也展示了所附属控制门变量的位置。

拥有了关键计算层的实现，便可以搭建模型网络。这一部分在 models 里面实现。此处以 VGG 模型进行展示，代码在 models/vgg.py 中。

```python
from layer import ConvBNReLU
import torch.nn as nn

cfgs = {                                    # 各标准模型通道数目配置
    'vgg16_bn': [
        64, 64, 128, 128,
        256, 256, 256,
        512, 512, 512,
        512, 512, 512
    ],
    'vgg19_bn': [
        64, 64, 128, 128,
        256, 256, 256, 256,
        512, 512, 512, 512,
        512, 512, 512, 512
    ]
}

maxpool_loc = {                             # 各模型 MaxPool 位置
    '16': [1, 3, 6, 9],
    '19': [1, 3, 7, 11]
}

class VGG(nn.Module):
    def __init__(self, depth, cfg, num_classes=10):
        super(VGG, self).__init__()
        self.depth = depth
```

```python
        self.cfg = cfg
        self.features = self._make_layers()
        self.avgpool = nn.AvgPool2d(2)
        self.classifier = nn.Linear(cfg[-1], num_classes)

    def _make_layers(self):
        maxpool_idx = maxpool_loc[str(self.depth)]
        layers = []
        in_channel = 3
        for i, n in enumerate(self.cfg):  # 根据配置情况，构建模型
            layers.append(
                ConvBNReLU(in_channel, n)
            )
            if i in maxpool_idx:                    # 添加 MaxPool 层
                layers.append(
                    nn.MaxPool2d(2, 2)
                )
            in_channel = n                          # 前一层输出通道数为下一层输入通道数

        return nn.Sequential(*layers)

    def forward(self, x):
        out = self.features(x)
        out = self.avgpool(out)
        out = out.view(out.size(0), -1)
        out = self.classifier(out)
        return out

def vgg16_bn(cfg=None, num_classes=10):  # 当无特定配置时，默认使用标准模型
    if cfg is None:
        cfg = cfgs['vgg16_bn']
    return VGG(16, cfg, num_classes)

def vgg19_bn(cfg=None, num_classes=10):
    if cfg is None:
        cfg = cfgs['vgg19_bn']
    return VGG(19, cfg, num_classes)
```

可以看出以上的实现方式中，我们既提供了默写的标准模型配置，又可以允许自定义配置来构建模型。这在构建剪枝模型时十分方便。并且在模型这个整体层面，并没有出现像 layer 中的 gated_forward 函数，因为我们可以通过将每个 layer 替换为控制门调节的前向输出，从而实现整体模型 forward 函数也被改变的效果。与之类似，ResNet 模型实现在 models/resnet.py 中：

```
from layer import ConvBNReLU, BasicBlock
import torch.nn as nn

cfgs = {                                          # 各标准模型配置
```

```python
    'resnet20': [16]*3 + [32]*3 + [64]*3,
    'resnet32': [16]*5 + [32]*5 + [64]*5,
    'resnet44': [16]*7 + [32]*7 + [64]*7,
    'resnet56': [16]*9 + [32]*9 + [64]*9,
}

stride2_loc = {                                     # 各标准模型 stride=2 位置处
    '20': [3, 6],
    '32': [5, 10],
    '44': [7, 14],
    '56': [9, 18]
}

class ResNet(nn.Module):
    def __init__(self, depth, cfg, num_classes=10):
        super(ResNet, self).__init__()
        self.depth = depth
        self.cfg = cfg
        self.conv = ConvBNReLU(3, 16)
        self.in_channel = 16
        self.features = self._make_layers()
        self.avgpool = nn.AvgPool2d(8)
        self.classifier = nn.Linear(
            self.in_channel, num_classes
        )

    def _make_layers(self):
        layers = []
        stride2_idx = stride2_loc[str(self.depth)]
        for i, n in enumerate(self.cfg):
            stride = 1                              # 正常情况下 stride=1
            out_channel = self.in_channel
            if i in stride2_idx:                    # 如果该层 stride 应为 2
                stride = 2
                out_channel = 2*self.in_channel     # 则输出通道数目扩大一倍

            layers.append(
                BasicBlock(
                    self.in_channel, n,
                    out_channel, stride=stride
                )
            )
            self.in_channel = out_channel           # 输出通道数为下一层输入数

        return nn.Sequential(*layers)

    def forward(self, x):
        out = self.conv(x)
```

```
            out = self.features(out)
            out = self.avgpool(out)
            out = out.view(out.size(0), -1)
            out = self.classifier(out)
            return out

    def resnet20(cfg=None, num_classes=10):              # 各标准模型的默认实现
        if cfg is None:
            cfg = cfgs['resnet20']
        return ResNet(20, cfg, num_classes)

    def resnet32(cfg=None, num_classes=10):
        if cfg is None:
            cfg = cfgs['resnet32']
        return ResNet(32, cfg, num_classes)

    def resnet44(cfg=None, num_classes=10):
        if cfg is None:
            cfg = cfgs['resnet44']
        return ResNet(44, cfg, num_classes)

    def resnet56(cfg=None, num_classes=10):
        if cfg is None:
            cfg = cfgs['resnet56']
        return ResNet(56, cfg, num_classes)
```

## 6.3.2 剪枝器实现

接下来介绍剪枝器的实现,相关代码在 pruner.py 中。这个类接受某种模型作为输入,并在指定待关联控制门的计算层类型后,负责将模型内部各计算层控制门变量初始化,转换前向运算过程等操作。而且还可以收集整体控制门变量,并根据某预定剪枝率进行裁剪。pruner.py 中还包含了关于控制门变量正则项的实现。因为在实际的训练过程中,需要添加对于控制门变量的正则项,以促进其具有某种性质,例如稀疏性。此处我们只实现了最简单的 $\ell_n$-范数正则项,且常用 $\ell_1$-范数以促进稀疏。具体代码如下:

```
import torch
import types


class Pruner(object):
    def __init__(self, model, module_type,
                 prune_rate=0.0, regularizer=None):
        self.model = model
        self.module_type = module_type
        self.prune_rate = prune_rate
        self.regularizer = regularizer
```

```python
        self.gates_params = []
        self.masks = []

    def init_gates(self):                        # 初始化控制门变量
        for m in self.model.modules():
            if m.__class__.__name__ == self.module_type:
                m.init_gates()

    def collect_gates(self):                     # 收集所有控制门变量
        for m in self.model.modules():
            if m.__class__.__name__ == self.module_type:
                self.gates_params.extend(m.get_gates())

    def replace_forward(self):                   # 替换前向操作
        for m in self.model.modules():
            if m.__class__.__name__ == self.module_type:
                m.forward = types.MethodType(
                    m.gated_forward, m
                )

    def gates_loss(self):                        # 获取施加于控制门变量的损失函数值
        return self.regularizer(self.gates_params)

    def calculate_mask(self):                    # 根据剪枝率计算删除掩码
        gates_lens = [len(gate) for gate in self.gates_params]
        all_gates = torch.cat(self.gates_params).abs()
        # 计算保留控制门的数量
        keep_gate_num = int(all_gates.numel() * (1 - self.prune_rate))
        # 获取保留控制门的位置索引
        keep_idx = all_gates.topk(keep_gate_num, 0)[1]
        masks = torch.zeros_like(all_gates)
        # 设置 0/1 掩码
        masks[keep_idx] = 1.0
        self.masks = torch.split_with_sizes(masks, gates_lens)

    def zerout_gates(self):                      # 根据掩码将控制门置零
        for i in range(len(self.gates_params)):
            self.gates_params[i].data.mul_(
                self.masks[i]
            )

    def export_pruned_cfg(self):                 # 导出每层保留通道数目
        cfg = []
        for m in self.masks:
            keep_num = max(m.sum().long().item(), 1)
            cfg.append(keep_num)
        return cfg
```

```
class LnRegularizer(object):                    # 基于Ln-范数的正则项
    def __init__(self, order=1):
        self.order = order

    def __call__(self, gates):
        all_gates = torch.cat(gates)
        ln_norm = torch.norm(all_gates, p=self.order) / len(all_gates)
        return ln_norm
```

## 6.3.3 学习控制门变量

接下来是学习控制门变量的 learn_gates.py 文件。包含了如何根据不同设定学习每层控制门变量。这里我们待实验的三种方法 L1-norm、Network Slimming 和 Soft Filter Pruning 在训练时有不同设置。L1-norm 和 Network Slimming 相似，都是在一起训练权重参数和控制门变量。区别在于前者并不对控制门变量施加正则项，而后者则是添加了 L1 范数。注意这里的 L1-norm 方法指的是在训练结束后，利用模型权重参数的 L1-norm 进行排序而剪枝，在训练过程中并未对控制门变量有任何约束。而此处我们这里添加了控制门变量用以直接学习得到权重重要性。这与原方法处理上稍有区别。而对于 Soft Filter Pruning，则是在每次迭代更新控制门变量后，需要将部分控制门直接抑制为0，这就需要使用 Pruner 类中的 zerout_gates 方法了。具体代码实现如下：

```
from torchvision import transforms, datasets
from pruner import Pruner, LnRegularizer
import torch.nn.functional as F
import numpy as np
import torch
import argparse
import os

import models
import misc

print = misc.logger.info

parser = argparse.ArgumentParser()              # 传递参数设置
parser.add_argument('--gpu', default='0', type=str)
parser.add_argument('--dataset', default='cifar10', type=str)
parser.add_argument('--arch', '-a', default='vgg16_bn', type=str)
parser.add_argument('--lr', default=0.1, type=float)
parser.add_argument('--mm', default=0.9, type=float)
parser.add_argument('--wd', default=1e-4, type=float)
parser.add_argument('--lambd', default=1e-3, type=float)
parser.add_argument('--epochs', default=160, type=int)
parser.add_argument('--log_interval', default=100, type=int)
```

```python
parser.add_argument('--train_batch_size', default=128, type=int)
parser.add_argument('--seed', default=None, type=int)
parser.add_argument('--prune_method', '-m',
                    default='slimming', type=str,
                    choices=['l1norm', 'slimming', 'softfilter'])
parser.add_argument('--module_type', '-t',
                    default='ConvBNReLU', type=str)
parser.add_argument('--prune_rate', '-p', default=0.0, type=float)

args = parser.parse_args()
args.seed = misc.set_seed(args.seed)           # 设置随机种子

args.device = 'cuda'
os.environ['CUDA_VISIBLE_DEVICES'] = args.gpu

args.num_classes = 10 if args.dataset == 'cifar10' else 100

args.logdir = '%s-%s/%s-learn_gates-%d' % (
    args.dataset, args.arch, args.prune_method, args.seed
)

misc.prepare_logging(args)

print('==> Preparing data..')

if args.dataset == 'cifar10':                  # 构建数据集
    transform_train = transforms.Compose([
        transforms.RandomCrop(32, padding=4),
        transforms.RandomHorizontalFlip(),
        transforms.ToTensor(),
        transforms.Normalize((0.4914, 0.4822, 0.4465),
                             (0.2023, 0.1994, 0.2010)),
    ])

    transform_val = transforms.Compose([
        transforms.ToTensor(),
        transforms.Normalize((0.4914, 0.4822, 0.4465),
                             (0.2023, 0.1994, 0.2010)),
    ])

    trainset = datasets.CIFAR10(
        root='./data/cifar10', train=True,
        transform=transform_train
    )
    trainloader = torch.utils.data.DataLoader(
        trainset, batch_size=args.train_batch_size,
        shuffle=True, num_workers=2
    )
```

```python
    valset = datasets.CIFAR10(
        root='./data/cifar10', train=False,
        transform=transform_val
    )
    valloader = torch.utils.data.DataLoader(
        valset, batch_size=100,
        shuffle=False, num_workers=2
    )
print('==> Initializing model...')              # 初始化模型
model = models.__dict__[args.arch](num_classes=args.num_classes)

print('==> Transforming model...')              # 初始化正则项
regularizer = LnRegularizer(order=1)

pruner = Pruner(                                # 构建剪枝器
    model, args.module_type,
    args.prune_rate, regularizer
)
pruner.init_gates()                             # 初始化控制门变量
pruner.collect_gates()                          # 收集控制门变量
pruner.replace_forward()                        # 替换利用控制门调节的前向过程

model = model.to(args.device)

optimizer = torch.optim.SGD(                    # 构建优化器
    model.parameters(),
    lr=args.lr, momentum=args.mm,
    weight_decay=args.wd
)

scheduler = torch.optim.lr_scheduler.MultiStepLR(  # 学习率调整器
    optimizer, milestones=[80, 120], gamma=0.1
)

def train(epoch):
    model.train()
    for i, (data, target) in enumerate(trainloader):
        data = data.to(args.device)
        target = target.to(args.device)

        # 前向反向过程
        optimizer.zero_grad()
        output = model(data)
        loss_ce = F.cross_entropy(output, target)
        loss_reg = args.lambd * pruner.gates_loss()
        loss = loss_ce + loss_reg
```

```python
            loss.backward()
            optimizer.step()

            if args.prune_method == 'softfilter':    # 计算掩码并置零
                pruner.calculate_mask()
                pruner.zerout_gates()


            if i % args.log_interval == 0:           # 输出信息
                all_gates = torch.cat(pruner.gates_params)
                mean_gate = all_gates.mean()
                acc = (output.max(1)[1] == target).float().mean()

                print('Train Epoch: %d [%d/%d]\t'
                    'Loss: %.4f, Loss_CE: %.4f, Loss_REG: %.4f, '
                    'Mean gate: %.4f, Accuracy: %.4f' % (
                    epoch, i, len(trainloader),
                    loss.item(), loss_ce.item(), loss_reg.item(),
                    mean_gate.item(), acc.item()
                ))

def test(epoch):                                     # 测试阶段
    model.eval()
    test_loss_ce = []
    correct = 0
    with torch.no_grad():
        for data, target in valloader:

            data, target = data.to(args.device), target.to(args.device)
            output = model(data)

            test_loss_ce.append(F.cross_entropy(output, target).item())

            pred = output.max(1)[1]
            correct += (pred == target).float().sum().item()

    acc = correct / len(valloader.dataset)
    print('Test Epoch: %d, Loss_CE: %.4f, '
        'Accuracy: %.4f\n' % (
        epoch, np.mean(test_loss_ce), acc
    ))

for epoch in range(args.epochs):
    train(epoch)
    test(epoch)
    torch.save(
        model.state_dict(),
        os.path.join(args.logdir, 'checkpoint.pth')
```

```
    )
    scheduler.step()
```

以上代码展示了如何学习控制门变量。例如对于使用 L1-norm 策略学习 VGG16 模型控制门变量，则运行以下命令：

```
python learn_gates.py -a vgg16_bn -t ConvBNReLU \
    -m l1norm --lambd 0.0 --seed 1234 --gpu 0
```

这里对该次实验设置随机种子，以便于多次重复性比较。在 misc.py 中包含了设置随机种子的方法，同时设置了 Python、numpy 和 torch 的随机性。

```
def set_seed(seed=None):
    if seed is None:
        seed = random.randint(0, 9999)
    np.random.seed(seed)
    torch.manual_seed(seed)
    torch.cuda.manual_seed_all(seed)
    torch.backends.cudnn.deterministic = True
    return seed
```

### 6.3.4 剪枝模型

在以上学习控制门变量结束后，便是根据之前所示的二分法进行剪枝结构搜索。具体代码在 prune_gates.py 中：

```
from pruner import Pruner
import argparse
import torch
import models
import os
import ptflops
import misc


parser = argparse.ArgumentParser()                  # 参数设置
parser.add_argument('--gpu', default='0', type=str)
parser.add_argument('--dataset', default='cifar10', type=str)
parser.add_argument('--arch', '-a', default='vgg16_bn', type=str)
parser.add_argument('--seed', default=None, type=int)
parser.add_argument('--prune_method', '-m',    # 学习控制门变量的方法
            default='slimming', type=str,
            choices=['l1norm', 'slimming', 'softfilter'])
parser.add_argument('--module_type', '-t',
            default='ConvBNReLU', type=str)
parser.add_argument('--prune_criterion', '-c', # 剪枝准则
            default='params', type=str,
            choices=['params', 'macs'])
parser.add_argument('--target_prune_rate', '-r', # 预定剪枝率
            default=0.5, type=float)
parser.add_argument('--max_prune_iter', '-i',  # 最多迭代次数
```

```python
                    default=10, type=int)
parser.add_argument('--eps', default=0.01, type=float)

args = parser.parse_args()
load_path = 'logs/%s-%s/%s-learn_gates-%d' % (
    args.dataset, args.arch, args.prune_method, args.seed
)
os.environ['CUDA_VISIBLE_DEVICES'] = args.gpu

args.num_classes = 10 if args.dataset == 'cifar10' else 100

model = models.__dict__[args.arch](num_classes=args.num_classes)
orig_macs, orig_params = ptflops.get_model_complexity_info(  # 原模型信息
    model, (3, 32, 32), print_per_layer_stat=False,
    verbose=False, as_strings=False
)
print('full model %s: %.2f(M) macs, %.2f(M) params' % (
    args.arch, orig_macs/1e6, orig_params/1e6
))

pruner = Pruner(model, args.module_type)
pruner.init_gates()                                          # 初始化控制门变量

model.load_state_dict(                                       # 加载已训练好的控制门变量
    torch.load(os.path.join(load_path, 'checkpoint.pth'))
)
pruner.collect_gates()

start_prune_rate = 0
end_prune_rate = 1

for j in range(args.max_prune_iter):                         # 利用二分法迭代进行剪枝
    cur_prune_rate = (start_prune_rate + end_prune_rate) / 2
    pruner.prune_rate = cur_prune_rate
    pruner.calculate_mask()                                  # 计算掩码
    pruned_model_cfg = pruner.export_pruned_cfg()            # 输出剪枝模型设置

    pruned_model = models.__dict__[args.arch](               # 构建剪枝模型结构
        pruned_model_cfg, num_classes=args.num_classes
    )
    prune_macs, prune_params = ptflops.get_model_complexity_info(
        pruned_model, (3, 32, 32), print_per_layer_stat=False,
        verbose=False, as_strings=False                      # 计算剪枝模型信息
    )
    if args.prune_criterion == 'params':                     # 计算实际剪枝率
        actual_prune_rate = 1 - prune_params / orig_params
    else:
        actual_prune_rate = 1 - prune_macs / orig_macs
```

```
    print('Iter %d, start %.2f, end %.2f, prune rate = %.4f' % (
        j, start_prune_rate, end_prune_rate, actual_prune_rate
    ))

    if abs(actual_prune_rate - args.target_prune_rate) <= args.eps:
        print('Successfully reach the target prune rate '
            'with %.2f(M) macs, %.2f(M) params' % (
            prune_macs/1e6, prune_params/1e6
        ))
        break                                        # 如果到达预定要求,停止

    if actual_prune_rate > args.target_prune_rate:   # 调整阈值进行下一次迭代
        end_prune_rate = cur_prune_rate
    else:
        start_prune_rate = cur_prune_rate

misc.dump_pickle(
    pruned_model_cfg,
    os.path.join(
        load_path, 'pruned_model_cfg-%.2f%s.pkl' % (
        args.target_prune_rate, args.prune_criterion
        )
    )
)
```

以上代码中,我们利用 prune_criterion 参数设置,是根据参数量剪枝还是计算量剪枝。默认情况下利用计算量剪枝,因为这更能实际反映出模型计算消耗。

### 6.3.5 训练模型

最后便是利用得到的剪枝模型结构设置,从头开始训练剪枝网络。其代码包含在 train_model.py 中:

```
from torchvision import transforms, datasets
import torch.nn.functional as F
import numpy as np
import torch
import argparse
import os

import models
import misc

print = misc.logger.info

parser = argparse.ArgumentParser()
```

```python
parser.add_argument('--gpu', default='0', type=str)
parser.add_argument('--dataset', default='cifar10', type=str)
parser.add_argument('--arch', '-a', default='vgg16_bn', type=str)
parser.add_argument('--lr', default=0.1, type=float)
parser.add_argument('--mm', default=0.9, type=float)
parser.add_argument('--wd', default=1e-4, type=float)
parser.add_argument('--epochs', default=160, type=int)
parser.add_argument('--log_interval', default=100, type=int)
parser.add_argument('--train_batch_size', default=128, type=int)
parser.add_argument('--prune_method', '-m',
            default='slimming', type=str,
            choices=['l1norm', 'slimming', 'softfilter'])
parser.add_argument('--prune_criterion', '-c',
            default='params', type=str,
            choices=['params', 'macs'])
parser.add_argument('--target_prune_rate', '-r',
            default=0.5, type=float)
parser.add_argument('--seed', default=None, type=int)
parser.add_argument('--baseline', action='store_true') # 是否训练基线模型
parser.add_argument('--ScratchB', action='store_true') # 是否使用 ScratchB

args = parser.parse_args()
args.seed = misc.set_seed(args.seed)

if args.ScratchB:                                    # 如果使用 ScratchB, 则延长训练时长
    args.epochs = int(1 / (1 - args.target_prune_rate) * args.epochs)

args.num_classes = 10 if args.dataset == 'cifar10' else 100

args.device = 'cuda'
os.environ['CUDA_VISIBLE_DEVICES'] = args.gpu

if args.baseline:
    args.logdir = '%s-%s/baseline-%d' % (
        args.dataset, args.arch, args.seed
    )
else:
    args.logdir = '%s-%s/%s-train_model-%.2f%s-%d' % (
        args.dataset, args.arch,
        args.prune_method, args.target_prune_rate,
        args.prune_criterion, args.seed
    )
    if args.ScratchB:
        args.logdir += '-ScratchB'

misc.prepare_logging(args)
```

```python
print('==> Preparing data..')

if args.dataset == 'cifar10':
    transform_train = transforms.Compose([
        transforms.RandomCrop(32, padding=4),
        transforms.RandomHorizontalFlip(),
        transforms.ToTensor(),
        transforms.Normalize((0.4914, 0.4822, 0.4465),
                             (0.2023, 0.1994, 0.2010)),
    ])

    transform_val = transforms.Compose([
        transforms.ToTensor(),
        transforms.Normalize((0.4914, 0.4822, 0.4465),
                             (0.2023, 0.1994, 0.2010)),
    ])

    trainset = datasets.CIFAR10(
        root='./data/cifar10', train=True,
        transform=transform_train
    )
    trainloader = torch.utils.data.DataLoader(
        trainset, batch_size=args.train_batch_size,
        shuffle=True, num_workers=2
    )

    valset = datasets.CIFAR10(
        root='./data/cifar10', train=False,
        transform=transform_val
    )
    valloader = torch.utils.data.DataLoader(
        valset, batch_size=100,
        shuffle=False, num_workers=2
    )

print('==> Initializing model...')
if args.baseline:                              # 训练基线模型，则使用标准设置
    model = models.__dict__[args.arch](None, args.num_classes)
else:                                          # 否则加载剪枝模型设置
    pruned_model_cfg = misc.load_pickle(
        'logs/%s-%s/%s-learn_gates-%d/pruned_model_cfg-%.2f%s.pkl' % (
            args.dataset, args.arch, args.prune_method, args.seed,
            args.target_prune_rate, args.prune_criterion
        )
    )
    model = models.__dict__[args.arch](pruned_model_cfg, args.num_classes)
```

```python
model = model.to(args.device)

optimizer = torch.optim.SGD(
    model.parameters(),
    lr=args.lr, momentum=args.mm,
    weight_decay=args.wd
)
scheduler = torch.optim.lr_scheduler.MultiStepLR(
    optimizer, milestones=[
        int(args.epochs * 0.5), int(args.epochs * 0.75)
    ], gamma=0.1
)


def train(epoch):
    model.train()
    for i, (data, target) in enumerate(trainloader):
        data = data.to(args.device)
        target = target.to(args.device)

        optimizer.zero_grad()
        output = model(data)
        loss = F.cross_entropy(output, target)
        loss.backward()
        optimizer.step()
        pred = output.max(1)[1]
        acc = (pred == target).float().mean()

        if i % args.log_interval == 0:
            print('Train Epoch: {} [{}/{}]'
                  '\tLoss: {:.6f}, Accuracy: {:.4f}'.format(
                epoch, i, len(trainloader),
                loss.item(), acc.item()
            ))

def test(epoch):
    model.eval()
    test_loss_ce = []
    correct = 0
    with torch.no_grad():
        for data, target in valloader:

            data, target = data.to(args.device), target.to(args.device)
            output = model(data)

            test_loss_ce.append(F.cross_entropy(output, target).item())
```

```
                pred = output.max(1)[1]
                correct += (pred == target).float().sum().item()

        acc = correct / len(valloader.dataset)
        print('Test Epoch: %d, Loss_CE: %.4f, '
            'Accuracy: %.4f\n' % (
          epoch, np.mean(test_loss_ce), acc
        ))

for epoch in range(args.epochs):
    train(epoch)
    test(epoch)
    torch.save(
        model.state_dict(),
        os.path.join(args.logdir, 'checkpoint.pth')
    )
    scheduler.step()
```

在这份代码中，我们既包含了训练剪枝模型的流程，也包含了训练基线模型的流程。二者唯一的差别即在于初始化模型架构的设置。同时也包含了是否使用 Scratch-B 训练流程，如果开启则延长训练周期时长。

## 6.3.6 规模化启动训练任务

编写完以上代码之后，我们就可以进行实验。例如对 VGG16 模型采取 Soft Filter Pruning 方法进行剪枝，每次迭代删减 50%通道，并最终使得剪枝模型减少原模型 50%的 MACs，训练时采用 Scratch-B 策略，设置随机种子为 1234 的实验，则可以运行如下代码：

```
python learn_gates.py -a vgg16_bn -t ConvBNReLU \
    -m softfilter -p 0.5 --seed 1234 --gpu 0
python prune_gates.py -a vgg16_bn -t ConvBNReLU \
    -m softfilter -c macs -r 0.5 --seed 1234 --gpu 0
- python train_model.py -a vgg16_bn -m softfilter \
    -c macs -r 0.5 --seed 1234 --ScratchB --gpu 0
```

如果运行正常则会输出以下结果，首先是 learn_gates.py 的输出 log：

```
[22:51:56.728] =================FLAGS==================
[22:51:56.728] gpu: 0
[22:51:56.728] dataset: cifar10
[22:51:56.728] arch: vgg16_bn
[22:51:56.728] lr: 0.1
[22:51:56.728] mm: 0.9
[22:51:56.728] wd: 0.0001
[22:51:56.728] lambd: 0.001
```

```
[22:51:56.728] epochs: 160
[22:51:56.728] log_interval: 100
[22:51:56.729] train_batch_size: 128
[22:51:56.729] seed: 1234
[22:51:56.729] prune_method: softfilter
[22:51:56.729] module_type: ConvBNReLU
[22:51:56.729] prune_rate: 0.5
[22:51:56.729] device: cuda
[22:51:56.729] num_classes: 10
[22:51:56.729] logdir: ./logs/cifar10-vgg16_bn/softfilter-learn_gates-1234
[22:51:56.729] ==========================================
[22:51:56.729] ==> Preparing data..
[22:51:57.782] ==> Initializing model...
[22:51:57.873] ==> Transforming model...
[22:52:01.326] Train Epoch: 0 [0/391]   Loss: 2.3416, Loss_CE: 2.3406,
Loss_REG: 0.0010, Mean gate: 0.5001, Accuracy: 0.0938
[22:52:06.184] Train Epoch: 0 [100/391] Loss: 1.7338, Loss_CE: 1.7333,
Loss_REG: 0.0005, Mean gate: 0.4941, Accuracy: 0.2656
[22:52:10.988] Train Epoch: 0 [200/391] Loss: 1.5789, Loss_CE: 1.5784,
Loss_REG: 0.0005, Mean gate: 0.4879, Accuracy: 0.3203
[22:52:15.762] Train Epoch: 0 [300/391] Loss: 1.6017, Loss_CE: 1.6012,
Loss_REG: 0.0005, Mean gate: 0.4822, Accuracy: 0.4062
[22:52:21.533] Test Epoch: 0, Loss_CE: 1.4893, Accuracy: 0.4224
...
```

学习权重通道结束后，是 prune_gates.py 的输出信息。

```
full model vgg16_bn: 314.16(M) macs, 14.72(M) params
Iter 0, start 0.00, end 1.00, prune rate = 0.7485
Iter 1, start 0.00, end 0.50, prune rate = 0.5521
Iter 2, start 0.00, end 0.25, prune rate = 0.1633
Iter 3, start 0.12, end 0.25, prune rate = 0.2692
Iter 4, start 0.19, end 0.25, prune rate = 0.4063
Iter 5, start 0.22, end 0.25, prune rate = 0.4977
Successfully reach the target prune rate with 157.80(M) macs, 8.83(M) params
Dumping pickle object to logs/cifar10-vgg16_bn/softfilter-learn_gates-1234/pruned_model_cfg-0.50macs.pkl
```

最后，是训练剪枝模型 train_model.py 的输出信息。

```
[01:12:19.442] =================FLAGS==================
[01:12:19.442] gpu: 0
[01:12:19.442] dataset: cifar10
[01:12:19.442] arch: vgg16_bn
[01:12:19.442] lr: 0.1
[01:12:19.442] mm: 0.9
[01:12:19.442] wd: 0.0001
[01:12:19.442] epochs: 320
[01:12:19.442] log_interval: 100
```

```
[01:12:19.443] train_batch_size: 128
[01:12:19.443] prune_method: softfilter
[01:12:19.443] prune_criterion: macs
[01:12:19.443] target_prune_rate: 0.5
[01:12:19.443] seed: 1234
[01:12:19.443] baseline: False
[01:12:19.443] ScratchB: True
[01:12:19.443] num_classes: 10
[01:12:19.443] device: cuda
[01:12:19.443] logdir: ./logs/cifar10-vgg16_bn/softfilter-train_model-0.50macs-1234-ScratchB
[01:12:19.443] =====================================
[01:12:19.444] ==> Preparing data..
[01:12:20.508] ==> Initializing model...
[01:12:23.600] Train Epoch: 0 [0/391]    Loss: 2.305990, Accuracy: 0.1094
[01:12:26.473] Train Epoch: 0 [100/391]  Loss: 1.976020, Accuracy: 0.2422
[01:12:29.316] Train Epoch: 0 [200/391]  Loss: 1.525422, Accuracy: 0.4141
[01:12:32.169] Train Epoch: 0 [300/391]  Loss: 1.364206, Accuracy: 0.4922
[01:12:36.222] Test Epoch: 0, Loss_CE: 1.4284, Accuracy: 0.4795
...
```

因此以上便实现了剪枝模型的完整流程。但是当我们需要大规模多组参数重复实验时，这种命令行式的输入既费时麻烦，也不方便调度。这里推荐大家使用 tmux 并搭配 tmuxp 这个辅助管理器。首先 tmux 是终端复用器（Terminal Multiplexer），用于在远程服务器上管理多个终端窗口（Terminal Window）。有时会遇到训练程序代码时间较长的情况，此时如果直接关闭和远程服务器的连接，那么程序会终止。tmux 提供了一种持续会话窗口，可以在退出登录时，将程序在后台挂起依然运行。而再此连接服务器时，登录上次会话，会打开当时的运行窗口继续对话。同时 tmux 还提供了丰富的会话窗口管理方法，可以开启多个标签栏和分屏，方便多任务同时进行。

tmux 的网上教程很多[①]，这里简单介绍一些命令操作。

```
$ tmux                       # 启动一个会话窗口
$ tmux detach                # 关闭当前窗口
$ tmux ls                    # 显示当前已有窗口名称
$ tmux attach -t 0           # 重新接入编号为 0 的会话
$ tmux attach -t <name>      # 按名字连接某个会话
```

进入会话窗口中后，可以对一个窗口进行分割，形成多个窗格面板（panel）。此时可以使用一些快捷键：

```
Ctrl+b %                     # 划分左右两个窗格
Ctrl+b "                     # 划分上下两个窗格
Ctrl+b <方向键>              # 在不同窗格之间切换
```

---

① https://www.ruanyifeng.com/blog/2019/10/tmux.html

```
Ctrl+b c                    # 创建新的窗口
Ctrl+b n                    # 切换下一个窗口
```

还有更多的快捷键，大家可以逐渐学习。图 6.3 展示了一个 tmux 可以实现的窗口划分。

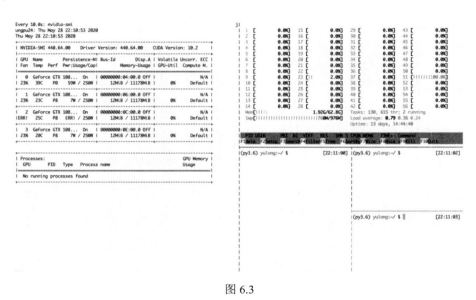

图 6.3

拥有了以上对于 tmux 的了解，我们可以进一步学习 tmuxp 这个窗口管理辅助工具。它可以通过编写各个窗口配置文件（YAML 格式），自动启动多个窗口及窗格。这可以方便我们管理大量的运行会话。安装方式也很简单，用 pip install tmuxp 即可。下面这个配置文件展示了基本使用方法：

```
session_name: 4-pane-split        # 会话名称
windows:
- window_name: dev window         # 窗口名称
  layout: tiled                   # 窗格布局模式
  shell_command_before:
    - cd ~/                       # 所有窗格均需运行的命令
  panes:
    - shell_command:              # 第一个窗格运行的命令
      - cd /var/log               # 按顺序运行多个命令
      - ls -al | grep \.log
    - echo second pane            # 第二个窗格运行的命令
    - echo third pane             # 第三个窗格
    - echo forth pane             # 第四个窗格
```

在 mysession.yaml 中编写完以上配置，然后在命令行中运行 tmuxp load mysession.yaml，则会得到如图 6.4 所示的窗口展示。

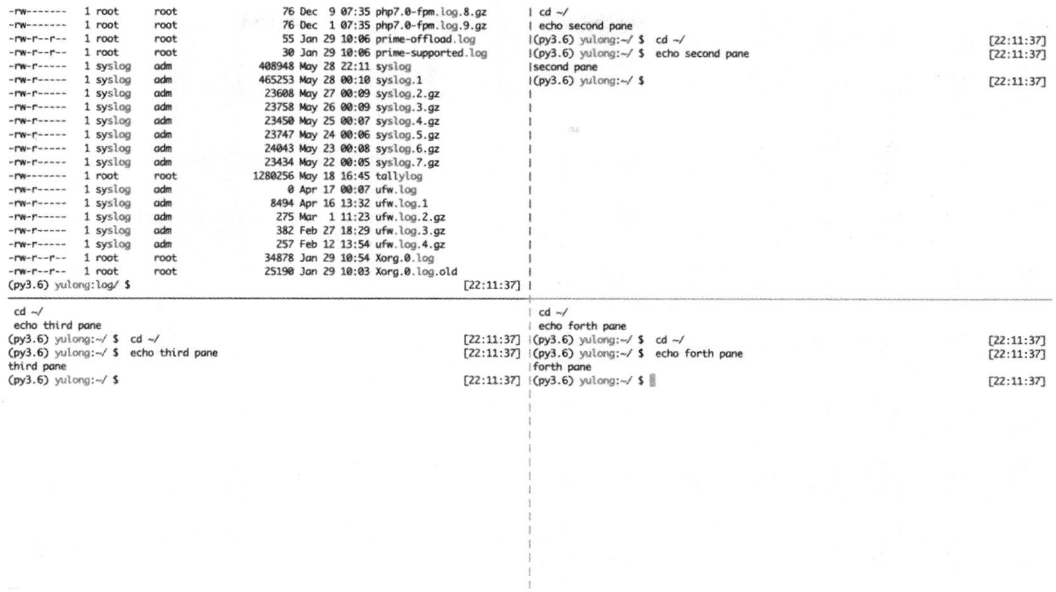

图 6.4

因此本次实验中，我们可以利用 tmuxp 来运行多组实验进行对比。例如以下便是设置随机种子为 1234，运行 L1-norm、Network Slimming 和 Soft Filter Pruning 剪枝方法，对 VGG16 和 VGG19 模型进行剪枝，并且按照 Scratch-B 训练流程进行训练。这六组实验可以分为三个窗口，每个窗口分为两个窗格，进行实验结果管理。

```
session_name: seed-1234
windows:
- layout: even-vertical
  window_name: l1norm
  panes:
  - shell_command:
    - python learn_gates.py -a vgg16_bn -t ConvBNReLU -m l1norm --lambd 0.0 --seed 1234 --gpu 0
    - python prune_gates.py -a vgg16_bn -t ConvBNReLU -m l1norm -c macs -r 0.5 --seed 1234 --gpu 0
    - python train_model.py -a vgg16_bn -m l1norm -c macs -r 0.5 --seed 1234 --ScratchB --gpu 0

  - shell_command:
    - python learn_gates.py -a vgg19_bn -t ConvBNReLU -m l1norm --lambd 0.0 --seed 1234 --gpu 1
    - python prune_gates.py -a vgg19_bn -t ConvBNReLU -m l1norm -c macs -r 0.5 --seed 1234 --gpu 1
    - python train_model.py -a vgg19_bn -m l1norm -c macs -r 0.5 --seed 1234 --ScratchB --gpu 1
```

```
      - layout: even-vertical
        window_name: slimming
        panes:
        - shell_command:
          - python learn_gates.py -a vgg16_bn -t ConvBNReLU -m slimming --seed
1234 --gpu 2
          - python prune_gates.py -a vgg16_bn -t ConvBNReLU -m slimming -c macs
-r 0.5 --seed 1234 --gpu 2
          - python train_model.py -a vgg16_bn -m slimming -c macs -r 0.5 --seed
1234 --ScratchB --gpu 2
        - shell_command:
          - python learn_gates.py -a vgg19_bn -t ConvBNReLU -m slimming --seed
1234 --gpu 3
          - python prune_gates.py -a vgg19_bn -t ConvBNReLU -m slimming -c macs
-r 0.5 --seed 1234 --gpu 3
          - python train_model.py -a vgg19_bn -m slimming -c macs -r 0.5 --seed
1234 --ScratchB --gpu 3

      - layout: even-vertical
        window_name: softfilter
        panes:
        - shell_command:
          - python learn_gates.py -a vgg16_bn -t ConvBNReLU -m softfilter -p 0.5
--seed 1234 --gpu 4
          - python prune_gates.py -a vgg16_bn -t ConvBNReLU -m softfilter -c macs
-r 0.5 --seed 1234 --gpu 4
          - python train_model.py -a vgg16_bn -m softfilter -c macs -r 0.5 --seed
1234 --ScratchB --gpu 4

        - shell_command:
          - python learn_gates.py -a vgg19_bn -t ConvBNReLU -m softfilter -p 0.5
--seed 1234 --gpu 5
          - python prune_gates.py -a vgg19_bn -t ConvBNReLU -m softfilter -c macs
-r 0.5 --seed 1234 --gpu 5
          - python train_model.py -a vgg19_bn -m softfilter -c macs -r 0.5 --seed
1234 --ScratchB --gpu 5
```

## 6.4 实验结果

在此次实验中，我们对比三种剪枝策略的剪枝模型最终性能。同时也包括了和基础完整模型性能的对比，以及使用 Scratch-B 训练策略的结果。所有模型均采用相同的训练策略，包括学习率、训练周期、学习率调整策略等等。这些参数都采用代码中的默认参数。所进行对比实验的模型包括了 VGG16、VGG19、ResNet20、ResNet56 这 4 种。所进行实验数据集为 CIFAR10，采取的数据增广策略也是标准的操作。为了更准确地进行比较，所有实验均使用 5 次不同随机种子进行完整实验流程，并最终报告 5 次实验结果的平均值。更多的模型剪枝结果和在其他数据集上的实验可以留待读者自行探索。

表 6.1 总结了实验结果。这里报告的是所有模型删减其原模型 50%计算量（MACs）后剪枝模型预测准确率。粗体显示的是基线完整模型的预测精度。斜粗体显示的是每个模型对应的最高剪枝策略结果。可以看出相比于不同训练方法，Scratch-B 训练方式可以有效提高最终模型训练精度。同时在诸多剪枝策略中，Soft Filter Pruning 方法在多数情况下可以取得最好效果。说明了在剪枝过程中进行多次删减会有助于控制门变量学习出更准确的各层通道的准确性

表 6.1　删减 50%计算量的剪枝模型准确率

|          | VGG16 | VGG19 | ResNet20 | ResNet56 |
|----------|-------|-------|----------|----------|
| baseline | 93.32 | 93.37 | 91.37    | 93.19    |
| l1norm   | 91.90 | 92.67 | 90.70    | 92.38    |
| slim     | 90.92 | 91.30 | 89.89    | 92.17    |
| soft     | 92.58 | 93.16 | 89.89    | 91.70    |
| l1norm-B | 92.57 | 93.18 | 90.61    | 92.53    |
| slim-B   | 92.46 | 92.22 | 90.66    | 92.70    |
| soft-B   | 93.33 | 93.41 | 90.46    | 92.93    |

同时我们也进行了另一组实验，要求所有方法删减原模型 70%的计算量，去观察各种方法能否在更大的剪枝率条件下依然实现较高的预测精度。表 6.2 总结了实验结果。其中的粗体和斜粗体所表达含义与表 6.1 相同。可以看出相比于删减 50%计算量时的剪枝模型准确率，当提高剪枝率之后，所得模型预测精度产生了下降。但是各种剪枝方法还是尽量避免了较大的精度损失。而且与之前观察结果相同，利用 Soft Filter Pruning 方法并结合 Scratch-B 训练策略可以在多数情况下达到最好的剪枝模型性能。

表 6.2　删减 70%计算量的剪枝模型准确率

|          | VGG16 | VGG19 | ResNet20 | ResNet56 |
|----------|-------|-------|----------|----------|
| baseline | 93.32 | 93.37 | 91.37    | 93.19    |
| l1norm   | 90.38 | 88.09 | 88.87    | 91.66    |
| slim     | 88.83 | 86.53 | 89.28    | 91.28    |
| soft     | 91.93 | 91.95 | 88.20    | 91.15    |
| l1norm-B | 91.89 | 92.05 | 89.66    | 91.80    |
| slim-B   | 91.51 | 92.29 | 88.64    | 92.15    |
| soft-B   | 92.89 | 92.67 | 89.46    | 92.09    |

除了以上数值展示，我们还可以观察不同剪枝策略所得到剪枝结构的不同。图 6.5~

图 6.8 分别展示了以上三种剪枝方法在 VGG16、VGG19、ResNet20 和 ResNet56 四种模型上所得到的各层保留通道数目。并且在图例中也注明了各个模型利用 Scratch-B 训练策略最终所得的模型预测精度。可以看出 L1-norm 和 Network Slimming 方法所得模型结构较为相似，Soft Filter Pruning 方法则更合理地保留了各层的通道数目。这也说明了在训练控制门变量过程中，如果能够加入剪枝过程，使得其预先适应通道删减情况，则有助于更合理地学习得到各层权重通道重要性，以便于得到更合理的模型结构。

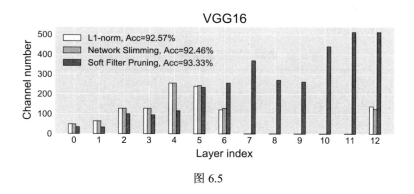

图 6.5

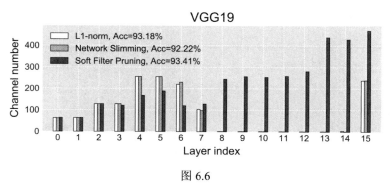

图 6.6

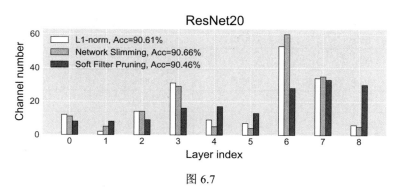

图 6.7

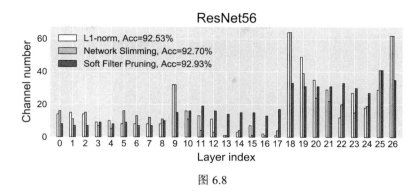

图 6.8

除此之外，我们还可以探究不同程度的剪枝率对于最终模型性能的影响。这里仅采取 Soft Filter Pruning 剪枝方法，并在 VGG16 和 ResNet56 模型上进行实验。对于每个模型，设置剪枝率从 0.1~0.9，间隔为 0.1。即要求删减原模型最低 10% 到最高 90% 的计算量。并且 Soft Filter Pruning 方法中在训练控制门变量过程中需要引入删减通道率，这个值也和剪枝率相同。即如果需要最终模型删减 70% 的计算量，则在学习控制门变量时每次迭代删除 70% 的权重通道。最终所有剪枝模型按照 Scratch-B 策略训练。而为了避免如删减 90% 计算量模型所需的训练周期过多（可能要达到 1600 个周期），此处截止最多训练周期为 600 个。

图 6.9 和图 6.10 便展示了以上的实验结果。可以看出所使用的剪枝策略在一定范围内均可以得到性能较好的剪枝模型。在剪枝率小于一定范围时，剪枝模型的预测性能甚至可以高于基线模型。当剪枝率大于某一程度之后，随着剪枝率提升，模型性能下降。这也说明了原始完整模型中存在着冗余结构与参数，可以被大幅缩减而不影响模型性能。

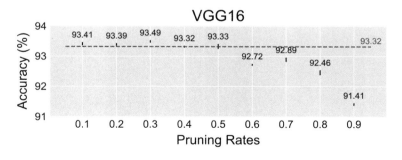

图 6.9

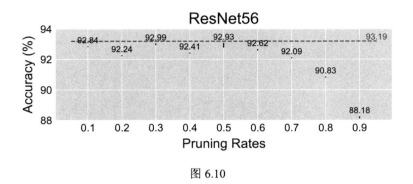

图 6.10

以上便是我们以网络剪枝这一问题进行的全面实验结果报告。文中并没有涉及到其他更多剪枝策略和更多数据集与模型。这部分内容有能力的读者可以自行探索实验。

# 参 考 文 献

[1] LECUN Y, BOSER B, DENKER J S, et al. Backpropagation applied to handwritten zip code recognition[J]. Neural Computation, 1989, 1(4): 541-551.

[2] HOCHREITER S, SCHMIDHUBER J. Long short-term memory[J]. Neural Computation, 1997, 9(8): 1735-1780.

[3] RUMELHART D E, HINTON G E, WILLIAMS R J. Learning representations by back-propagating errors[J]. Nature, 1986, 323(6088): 533-536.

[4] WALLACE R, STENTZ A, THORPE C, et al. First results in robot road-following[C]// Proceedings of the 9th International Joint Conference on Artificial Intelligence-Volume 2. 1985: 1089-1095.

[5] KRIZHEVSKY A, SUTSKEVER I, HINTON G E. Imagenet classification with deep convolutional neuralnetworks[C]// Advances in Neural Information Processing Systems 25. Curran Associates Inc., 2012: 1097-1105.

[6] SIMONYAN K, ZISSERMAN A. Very deep convolutional networks for large-scale image recognition[C]//International Conference on Learning Representations. ICLR Committee, 2015.

[7] SZEGEDY C, LIU W, JIA Y, et al. Going deeper with convolutions[C]//Proceedings of the IEEE Conference on Computer Vision and Pattern Recognition. IEEE, 2015: 1-9.

[8] HE K, ZHANG X, REN S, et al. Deep residual learning for image recognition[C]// Proceedings of the IEEE Conference on Computer Vision and Pattern Recognition. IEEE, 2016: 770-778.

[9] SILVER D, SCHRITTWIESER J, SIMONYAN K, et al. Mastering the game of go without human knowledge[J]. Nature, 2017, 550(7676):354-359.

[10] VASWANI A, SHAZEER N, PARMAR N, et al. Attention is all you need[C]// Advances in Neural Information Processing Systems 30. Curran Associates Inc., 2017: 5998-6008.

[11] DEVLIN J, CHANG M W, LEE K, et al. Bert: Pre-training of deep bidirectional transformers for language understanding[J]. arXiv preprint arXiv:1810.04805, 2018.

[12] ZOPH B, LE Q V. Neural architecture search with reinforcement learning[J]. arXiv preprint arXiv:1611.01578, 2016.

[13] COLLOBERT R, KAVUKCUOGLU K, FARABET C. Torch7: A matlab-like environment for machine learning[C]//BigLearn, NIPS workshop. 2011.

[14] BENGIO Y, LÉONARD N, COURVILLE A. Estimating or propagating gradients through stochastic neurons for conditional computation[J]. arXiv preprint arXiv:1308.3432, 2013.

[15] MADDISON C J, MNIH A, TEH Y W. The concrete distribution: A continuous relaxation of discrete random variables[C]//International Conference on Learning Representations. Toulon, France: ICLR Committee, 2017.

[16] JANG E, GU S, POOLE B. Categorical reparameterization with gumbel-softmax[C]//International Conference on Learning Representations. Toulon, France: ICLR Committee, 2017.

[17] SPRINGENBERG J T, DOSOVITSKIY A, BROX T, et al. Striving for simplicity: The all convolutional net[J]. arXiv preprint arXiv:1412.6806, 2014.

[18] SZEGEDY C, LIU W, JIA Y, et al. Going deeper with convolutions[C]//Proceedings of the IEEE Conference on Computer Vision and Pattern Recognition. IEEE, 2015: 1-9.

[19] HE K, ZHANG X, REN S, et al. Delving deep into rectifiers: Surpassing human-level performance on imagenet classification[C]//Proceedings of the IEEE International Conference on Computer Vision. Santiago, Chile: IEEE, 2015: 1026–1034.

[20] XIE S, GIRSHICK R B, DOLLÁR P, et al. Aggregated residual transformations for deep neural networks[C]//Proceedings of the IEEE Conference on Computer Vision and Pattern Recognition. Hawaii, USA: IEEE, 2017: 5987-5995.

[21] CHOLLET F. Xception: Deep learning with depthwise separable convolutions[C]//Proceedings of the IEEE Conference on Computer Vision and Pattern Recognition. Venice, Italy: IEEE, 2017: 1800-1807.

[22] SANDLER M, HOWARD A, ZHU M, et al. Mobilenetv2: Inverted residuals and linear bottlenecks [C]//Proceedings of the IEEE Conference on Computer Vision and Pattern Recognition. Salt Lake City, USA: IEEE, 2018: 4510-4520.

[23] ZHANG Z, LI J, SHAO W, et al. Differentiable learning-to-group channels via groupable convolutional neural networks[C]//Proceedings of the IEEE International Conference on Computer Vision. Seoul, Korea: IEEE, 2019: 3541-3550.

[24] ZHU J, PARK T, ISOLA P, et al. Unpaired image-to-image translation using cycle-consistent adversarial networks[C]//Proceedings of the IEEE International Conference on Computer Vision. Venice, Italy: IEEE, 2017: 2242-2251.

[25] HINTON G E, SRIVASTAVA N, KRIZHEVSKY A, et al. Improving neural networks by preventing coadaptation of feature detectors[J]. arXiv preprint arXiv:1207.0580, 2012.

[26] GIRSHICK R. Fast r-cnn[C]//Proceedings of the IEEE International Conference on Computer Vision. Santiago, Chile: IEEE, 2015: 1440 -1448.

[27] GRAVES A, FERNÁNDEZ S, GOMEZ F. Connectionist temporal classification: Labelling unsegmented sequence data with recurrent neural networks[C]//Proceedings of the International Conference on Machine Learning. Pittsburgh, USA: JMLR.org, 2006: 369-376.

[28] HOCHREITER S, SCHMIDHUBER J. Long short-term memory[J]. Neural Computation, 1997, 9(8): 1735–1780.

[29] JADERBERG M, SIMONYAN K, ZISSERMAN A, et al. Spatial transformer networks[M]//Advances in Neural Information Processing Systems 28. Montréal, Canada: Curran Associates, Inc., 2015: 2017-2025.

[30] SUTSKEVER I, MARTENS J, DAHL G, et al. On the importance of initialization and momentum in deep learning[C]//Proceedings of Machine Learning Research: volume 28 Proceedings of the 30th International Conference on Machine Learning. Atlanta, USA: PMLR, 2013: 1139-1147.

[31] KINGMA D P, BA J. Adam: A method for stochastic optimization[C]//International Conference on Learning Representations. Banff, Canada: ICLR Committee, 2014.

[32] LOSHCHILOV I, HUTTER F. Decoupled weight decay regularization[C]//International Conference on Learning Representations. New Orleans, USA: ICLR Committee, 2019.

[33] LOSHCHILOV I, HUTTER F. Sgdr: Stochastic gradient descent with warm restarts[C]//International Conference on Learning Representations. San Juan, Puerto Rico: ICLR Committee, 2016.

[34] HUBARA I, COURBARIAUX M, SOUDRY D, et al. Binarized neural networks[C]//Advances in Neural Information Processing Systems 29. Barcelona, Spain: Curran Associates, Inc., 2016: 4107-4115.

[35] PENNINGTON J, SOCHER R, MANNING C. Glove: Global vectors for word representation[C]//Proceedings of the 2014 Conference on Empirical Methods in Natural Language Processing (EMNLP). Doha, Qatar: Association for Computational Linguistics, 2014: 1532-1543.

[36] CUBUK E D, ZOPH B, MANE D, et al. Autoaugment: Learning augmentation policies from data [C]//Proceedings of the IEEE Conference on Computer Vision and Pattern Recognition. Long Beach, USA: IEEE, 2019: 113-123.

[37] TAN M, LE Q. Efficientnet: Rethinking model scaling for convolutional neural networks[C]//Proceedings of Machine Learning Research: volume 97 Proceedings of the 36th International Conference on Machine Learning. Long Beach, USA: PMLR, 2019: 6105-6114.

[38] TAN M, LE Q. Mixconv: Mixed depthwise convolutional kernels[C]//Proceedings of the British Machine Vision Conference (BMVC). London, UK: BMVA Press: 1-13.

[39] HOWARD A, SANDLER M, CHU G, et al. Searching for mobilenetv3[C]//Proceedings of the IEEE International Conference on Computer Vision. Seoul, Korea: IEEE, 2019: 1314-1324.

[40] MAATEN L, HINTON G. Visualizing data using t-SNE[J]. Journal of Machine Learning Research, 2008, 9(Nov):2579-2605.

[41] CAI H, ZHU L, HAN S. ProxylessNAS: Direct neural architecture search on target task and hardware[C]//International Conference on Learning Representations. New Orleans, USA: ICLRCommittee, 2019.

[42] GUO Z, ZHANG X, MU H, et al. Single path one-shot neural architecture search with uniform sampling[J]. arXiv preprint arXiv:1904.00420, 2019.

[43] DEVRIES T, TAYLOR G W. Improved regularization of convolutional neural networks with cutout [J]. arXiv preprint arXiv:1708.04552, 2017.

[44] SELVARAJU R R, COGSWELL M, DAS A, et al. Grad-cam: Visual explanations from deep networks via gradient-based localization[C]//Proceedings of the IEEE International Conference on Computer Vision. Venice, Italy: IEEE, 2017: 618-626.

[45] ZHOU B, KHOSLA A, LAPEDRIZA A, et al. Learning deep features for discriminative localization [C]//Proceedings of the IEEE Conference on Computer Vision and Pattern Recognition. Las Vegas, USA: IEEE, 2016: 2921-2929.

[46] SHANG W, SOHN K, ALMEIDA D, et al. Understanding and improving convolutional neural networks via concatenated rectified linear units[C]//Proceedings of the 33rd International Conference on International Conference on Machine Learning - Volume 48. New York, USA: JMLR.org, 2016: 2217–2225.

[47] ZAGORUYKO S, KOMODAKIS N. Diracnets: Training very deep neural networks without skipconnections[J]. arXiv preprint arXiv:1706.00388.

[48] SANDLER M, HOWARD A, ZHU M, et al. Mobilenetv2: Inverted residuals and linear bottlenecks [C]//Proceedings of the IEEE Conference on Computer Vision and Pattern Recognition. Salt Lake City, USA: IEEE, 2018: 4510-4520.

[49] SZEGEDY C, VANHOUCKE V, IOFFE S, et al. Rethinking the inception architecture for computer vision[C]//Proceedings of the IEEE Conference on Computer Vision and Pattern Recognition. Las Vegas, USA: IEEE, 2016: 2818-2826.

[50] GOYAL P, DOLLÁR P, GIRSHICK R B, et al. Accurate, large minibatch SGD: training imagenet in 1 hour[J]. arXiv preprint arXiv:1706.02677, 2017.

[51] TAN M, LE Q. Mixconv: Mixed depthwise convolutional kernels[C]//Proceedings of the British Machine Vision Conference (BMVC). London, UK: BMVA Press: 1-13.

[52] HAN S, MAO H, DALLY W J. Deep compression: Compressing deep neural networks with

pruning, trained quantization and huffman coding[C]//International Conference on Learning Representations. San Diego, USA: ICLR Committee, 2015.

[53] LECUN Y, DENKER J S, SOLLA S A. Optimal brain damage[C]//Proceedings of the 5th International Conference on Neural Information Processing Systems. Denver, USA: Curran Associates Inc., 1990: 598-605.

[54] HAN S, LIU X, MAO H, et al. EIE: Efficient inference engine on compressed deep neural network[C]//2016 ACM/IEEE 43rd Annual International Symposium on Computer Architecture (ISCA). Seoul, South Korea: IEEE, 2016: 243-254.

[55] LI H, KADAV A, DURDANOVIC I, et al. Pruning filters for efficient convnets[J]. arXiv preprint arXiv:1608.08710, 2016.

[56] LIU Z, LI J, SHEN Z, et al. Learning efficient convolutional networks through network slimming [C]//Proceedings of the IEEE International Conference on Computer Vision. Venice, Italy: IEEE, 2017: 2736-2744.

[57] HE Y, KANG G, DONG X, et al. Soft filter pruning for accelerating deep convolutional neural networks[C]//Proceedings of the 27th International Joint Conference on Artificial Intelligence. New Orleans, USA: AAAI Press, 2018: 2234-2240.

[58] TIBSHIRANI R. Regression shrinkage and selection via the lasso[J]. Journal of the Royal Statistical Society: Series B (Methodological), 1996, 58(1):267-288.

[59] HE Y, ZHANG X, SUN J. Channel pruning for accelerating very deep neural networks[C]//Proceedings of the IEEE International Conference on Computer Vision. Venice, Italy: IEEE, 2017: 1389-1397.

[60] LUO J H, WU J, LIN W. Thinet: A filter level pruning method for deep neural network compression[C]//Proceedings of the IEEE International Conference on Computer Vision. Venice, Italy: IEEE, 2017: 5058-5066.

[61] HE Y, LIN J, LIU Z, et al. Amc: Automl for model compression and acceleration on mobile devices[C]//Proceedings of the European Conference on Computer Vision. Munich, Germany: Springer, 2018: 784-800.

[62] LIU Z, SUN M, ZHOU T, et al. Rethinking the value of network pruning[C]//International Conference on Learning Representations. New Orleans, USA: ICLR Committee, 2019.